ADAPTATION TO DESERT ENVIRONMENT

> Θεὸς ἀεὶ γεωμετρεῖ
> Everywhere Nature works true to scale, and
> everything has its proper size accordingly.
> — D'ARCY W. THOMPSON

> *Le Désert*
> C'est l'horizon trompeur, c'est l'infini Désert,
> Où ni carte, ni plan, ni boussole ne sert,
> Où, de trombe en mirage, erre la caravane;
>
> Le royaume éternel du soleil et du vent,
> La frontière de sable où la tempête vanne:
> Pour ciel un brasier bleu, pour sol un flot mouvant!
> — ED. L. DE LAGARENNE
> *Alexandria,* 1897

ADAPTATION TO DESERT ENVIRONMENT

A STUDY ON THE JERBOA, RAT AND MAN

J. P. KIRMIZ, D.Sc.
*Physiological Research Laboratory,
Alexandria, Egypt, U.A.R.*

LONDON
BUTTERWORTHS
1962

ENGLAND:	BUTTERWORTH & CO. (PUBLISHERS) LTD. LONDON: 88 Kingsway, W.C.2
AFRICA:	BUTTERWORTH & CO. (AFRICA) LTD. DURBAN: 33–35 Beach Grove
AUSTRALIA:	BUTTERWORTH & CO. (AUSTRALIA) LTD. SYDNEY: 6–8 O'Connell Street MELBOURNE: 473 Bourke Street BRISBANE: 240 Queen Street
CANADA:	BUTTERWORTH & CO. (CANADA) LTD. TORONTO: 1367 Danforth Avenue, 6
NEW ZEALAND:	BUTTERWORTH & CO. (NEW ZEALAND) LTD. WELLINGTON: 49–51 Ballance Street AUCKLAND: 35 High Street
U.S.A.:	BUTTERWORTH INC. WASHINGTON, D.C.: 7235 Wisconsin Avenue, 14

Suggested UDC Number: 599·323·3: 591·1 + 591·5 (252)

©
Butterworth & Co. (Publishers) Ltd.
1962

Made and printed in Great Britain by
William Clowes and Sons, Limited, London and Beccles

INTRODUCTION

Living organisms possess the power of adaptation to their environment. For an organism to adapt it must adjust its internal responses to changing external conditions. Furthermore, different environments necessitate varying capacities of adaptation, and there are environments to which adaptation is not possible.

Both the adapting organism and the environment to which it adapts itself form the integral parts of a complex and dynamic system that tends towards a certain equilibrium. Life on earth is made possible by the physico-chemical conditions which favour its existence and by the modifications which living beings bring to their environment.

To explain the origin and development of species on earth, it is necessary to study the interaction and interdependence between living organisms and their environment. Survival is essentially a process of adaptation, and in their struggle for existence under changing environmental conditions living beings have developed adaptive mechanisms. It is one of the tasks of biological sciences to study the vital processes of adjustment that occur within the organism as a complex unit, and the adaptations that animals, in common with other living beings, are able to make to their environment.

Claude Bernard's dictum—'The essential condition for free life is the constancy of the internal milieu'—refers to the internal regulative and adaptive mechanisms developed by higher forms of life in the course of evolution, which assure them a certain degree of stability in an unstable and at times hostile environment.

The limits within which an animal can survive in a particular environment are determined by the functional responses of its internal milieu to the external conditions of stress. The individuals of each species are endowed with varying degrees of adaptive power. These individual differences play an important role in the struggle for survival.

Environmental stresses call for adaptive reactions, which not only transform the individual, both functionally and structurally, in the course of time, but also help the process of natural selection by favouring the survival of those individuals within the species which have acquired, genetically or by mutation, a higher power of adaptation.

As Cuénot (1951) stated, biological evolution presents the phenomenon of convergence, meaning that the resemblances among living beings are not only hereditary but also represent analogous reactions to analogous environmental conditions.

During recent years there has been a growing interest in the study of animal physiology under conditions of stress. The desert environment is considered to be an excellent field of investigation for this purpose, and significant research on Man has been undertaken between 1940 and 1946.

What adaptive and regulative mechanisms have the desert animals developed which enable them to withstand the extreme conditions of environmental strain in desert climates? Many questions remain to be answered in this new field of investigation—that is, the physiology of desert animals.

In fact, all studies related to life and conditions in the deserts are comparatively new. At present many scientists throughout the world are focusing their attention on the investigation of desert problems and the development of desert regions. Already there are desert institutes in several countries, and conferences on desert problems are being organized on an international scale. The interrelated problems of the desert cannot be solved by the isolated studies of a few branches of science alone. All the disciplines must investigate this field and correlate their findings before we may analyse its complex factors and their repercussions on living beings.

The purpose of this study is to discover how a desert animal like the jerboa (*Dipus aegyptius* or *Jaculus orientalis*) can adapt itself to its desert environment and, more precisely, how it regulates its temperature and energy needs in comparison with a rodent of temperate climates, the white rat. To answer this question we have proceeded as follows:

Firstly, we have studied the writings of different authors, geographers, ethnologists and biologists, concerning the desert, its climatic conditions, its flora and fauna. We have reviewed the essential findings, particularly during the last 20 years, on man's adaptation to desert climates.

Secondly, we have, through regular visits to the Egyptian Western Desert during different seasons, studied the jerboa's natural habitat and habits of living. We have completed our observations with valuable data collected from the bedouin, the dwellers of the desert, who are acquainted with certain characteristics of the way jerboas live.

Thirdly—and this is the object of the experimental part of our study—we have undertaken in the laboratory comparative research on the energy metabolism and thermoregulation of the jerboa and white rat. In planning our experimental methods we have taken into account the exigencies of the desert species, with a view to conserving the experimental animal in good condition.

We could not, with the limited resources for research at our disposal, examine all the aspects of the jerboa's energetics and thermoregulation. Nevertheless, we were able to collect precise data on the life of a hitherto little known desert species. We have also made an attempt to review the results of studies on man's adaptation to desert climate.

We trust that this study will arouse the interest of investigators in desert species other than the jerboa, and stimulate the continuation of studies on the jerboa and Man, because many problems remain unsolved.

Alexandria, Egypt, U.A.R.
January, 1962

JOHN P. KIRMIZ

ACKNOWLEDGEMENTS

The author is indebted to Professor E. Le Breton and Dr L. Dontcheff, of the Faculty of Science of Paris University, La Sorbonne. He thanks the authors whose writings he consulted and the friends who encouraged the preparation of this study. His special thanks are due to the Publishers.

CONTENTS

Introduction v

1. Adaptation of Biota to Desert Environment 1
   Introduction 1
   Hot deserts 2
   Macroclimate 3
   Microclimates and desert life 6
   Morphological and physiological adaptations 10
   Conclusions 13

2. Jerboa as a Desert Species 14
   Origin 14
   Distribution 15
   Description of the jerboa 16

3. The Jerboa's Habitat 21
   Desert environment 21
   The burrow and the jerboa's ecoclimate 24
   The jerboa's life 29
   Conclusions 30

4. The Jerboa in the Laboratory 32
   Nutrition 32
   The pseudo-burrow 32
   Rest and sleep 34
   Behaviour 35
   Identification 35
   Reproduction 36
   Conclusions 36

5. Comparative Growth of the Jerboa and Rat: Diet and Body Weight 37
   Growth 37
   Nutrition and body weight 39
   Conclusions 41

6. Effect of Diet on Excretion and Body Water Content . . 42
   Introduction 42
   Nutrition and excretion of the jerboa 42
   Effect of diet on excretion of jerboas and rats 44
   Water content of the body 45
   Spontaneous activity and nutrition 47
   Conclusions 48

7. Regulation of Body Temperature 50
   Body-temperature variations 50
   Body temperature of the jerboa and rat 50

CONTENTS

Zone of body-temperature regulation in jerboas on wet and dry diets	52
Comparison of zones of body-temperature regulation in jerboas and rats	53
Conclusions	55

8. Energy Metabolism in Relation to External Temperatures . . 57

Basal metabolism	57
Thermoregulation	58
Experiments	59
Results	60

9. Insensible Perspiration and Evaporative Water Loss in Relation to Thermal Environment 64

Introduction	64
Experiments	65
Results	66
Interpretation of results	69
Conclusions	70

10. Thermoregulation of the Jerboa and Rat 72

Heat balance and water balance	72
Thermoregulation	75
Conclusions	75

11. Man's Adaptation to Desert Climate 76

Climate and the human race	76
Man and the deserts	77
Physiological studies on effects of climatic elements	79
Hot desert climate	80
Heat balance	80
Water and salt balance	83
Sweating	85
Acclimatization	87
Desert hazards	88
Conclusions	90

12. Summary and Conclusions 91

Adaptation	91
Comparative study of rodents	91
The jerboa and its habitat	91
The jerboa in the laboratory	91
The jerboa compared with the rat	91
Effects of diet	92
Body temperature	92
Energy metabolism	92
Insensible perspiration	93
Thermoregulation	93
Man's adaptation to desert	94
General conclusions	94

CONTENTS

APPENDIXES

1. Technique 99
2. Body weight of jerboas 113
3. Dimensions of the jerboa 114
4. Temperature and humidity of the jerboa's habitat 115
5. Comparative growth of the jerboa and rat 116
   Jerboas 116
   White rats 117
6. Variation of the jerboa's body weight during growth and after maturity . 118
7. Effect of dry diet 119
   Effect of dry diet on body weight of adult jerboas . . . 119
   Body-weight variations of jerboas on dry diet 120
8. Effect of wet and dry diet 121
   Effect of alternating wet and dry diets on body weight of jerboas . 121
   Effects of wet and dry diets on the excretion of jerboas:
      Group I: jerboas habituated to wet diet for more than one year . 122
      Group II: jerboas habituated to dry diet for a period of more than
         one year 128
   Effect of temporary wet and dry diets on excretion of jerboas and white
      rats 130
9. Body temperature of the jerboa and white rat 132
   White rat (Wistar) 132
   Jerboas on wet diet 132
   Jerboas on dry diet 133
10. Zone of body-temperature regulation 134
    Jerboas on wet diet 134
    Jerboas on dry diet 134
    Zone of body-temperature regulation of jerboas and white rats:
       Jerboas 135
       White rats 136
11. Energy metabolism in relation to external temperature . . . 137
    Respiratory exchanges of jerboas on wet diet 137
    Respiratory exchanges of jerboas on dry diet 139
    Respiratory exchanges of white rats 140
12. Water loss of jerboas and rats 143
    Insensible perspiration and evaporative water loss of jerboas on wet diet 143
    Insensible perspiration and evaporative water loss of jerboas on dry diet 144
    Insensible perspiration and evaporative water loss of white rats . . 144
    Latent heat of evaporation in jerboas and rats in relation to total heat 145
    Evaporative water loss in relation to oxygen consumed . . . 147

BIBLIOGRAPHY 149

INDEX 159

1
ADAPTATION OF BIOTA TO DESERT ENVIRONMENT

INTRODUCTION

In recent times geographers have estimated that the total area of the earth is 510,101,235 km^2, of which 29·12 per cent represent land surface (continental area) and 70·88 per cent water surface (maritime area). Of the continental area, 57·9 per cent constitute fertile regions, 33·33 per cent steppes and 8·77 per cent, or about 13,000,000 km^2, deserts (*World Almanac*, 1956). But on the basis of the latest definition of the term 'desert region', it is reckoned that the existing desert and semi-desert lands vary between 18,500,000 and 25,000,000 km^2.

According to Unstead (1948), the arid and semi-arid regions are located on both sides of the Tropics of Cancer and Capricorn. They receive a great amount of insolation and heat, and little rain. They are in the dry, high-pressure belt, in the interior of the continents or on their western sides. The fertile river valleys and oases in these regions occupy only a small portion of the vast expanses of desert land. These desert regions are situated in two great belts, encircling the earth, approximately between latitudes 15° and 40° N and S of the equator. They are distributed, in order of area (in square kilometres), as follows:

Sahara, 6,734,000	Colorado, 518,000
Australia, 2,849,000	Gobi, 466,200
Turkestan, 2,331,000	Kalahari, 233,100
Arabia, 1,243,000	Thar, 191,660
Argentina, 1,036,000	Chile, 191,650

In order to gain a comparative impression of the world's desert areas, land surface and population by continent, *Table 1* was prepared on the basis of the estimates quoted in the *World Almanac* (1956) and the measurements made by Taylor (1927) from Koppen's map.

Table 1

World population, land area and deserts

Continent	Population		Land area*		Deserts	
	(*millions*)	%	(km^2)	%	(km^2)	%
Asia (incl. USSR)	1520	60·97	49,264,390	36·46	4,232,060	26·79
Europe	403·1	16·17	4,905,460	3·63	—	—
America	348	13·96	42,092,680	31·16	1,745,660	11·05
Africa	208	8·34	30,328,900	22·45	6,967,100	44·11
Australia	13·9	0·56	8,513,330	6·30	2,849,000	18·05
Total	2493	100	135,104,760	100	15,793,820	100

* Polar regions excluded.

This table clearly indicates that the African continent comprises 44·11 per cent of the world's desert regions, while its population accounts for only 8·34 per cent, spread over an area of about 22·45 per cent of the world's land surface.

Koppen's original definition of desert land—namely, 'any area where the annual rainfall in centimetres amounts to a number smaller than the mean annual temperature in degrees centigrade plus 16·5'—is now considered to be conservative. As the above estimates are based on Koppen's definition of desert, the actual arid and semi-arid regions of the world are much larger in extent, particularly from the bioclimatic point of view. In fact, the desert areas have no known boundaries. Moreover, desert conditions are continually changing: they are not static. An area which may qualify as 'desert' according to a given formula in one period may not do so exactly in another. This is why there seems to be no standard definition of what constitutes a desert country. Desert scholars differ on the question of criteria needed to differentiate a desert region from a non-desert one; and yet deserts exist, and scientists are at present trying to establish a desert terminology. What the future holds for desert-occupied countries one cannot tell, but surely vast stretches of the earth's land surface—approximately one fifth—lie unexplored and undeveloped, waiting for modern science and technology to reclaim and develop them for man's benefit.

HOT DESERTS

There are different types of desert. The simplest known classification is: (*a*) hot deserts, (*b*) temperate deserts and (*c*) cold deserts.

Among the elements of nature that operate in the formation of deserts are sea, rain, heat, cold, wind, aridity, edaphic conditions, geological eruptions and climatic changes. As rains fail and droughts occur, aridity increases; winds transfer dust and sand from place to place, and deserts advance. The history of the existing deserts remains to be written.

A hot desert environment—for example, the central Sahara—is marked by the following characteristics:

(1) Aridity, or lack of adequate water supply, which imposes varying degrees of limitation to the growth and development of plants and animals.

(2) Extremes of temperature—excessive diurnal, nocturnal and seasonal variations.

(3) Dryness of air, or low relative humidity.

(4) Fast evaporation.

(5) Limited and irregular rainfall from year to year, or no rainfall at all for several years, or sudden torrential showers followed by long periods of no rain whatsoever.

(6) Intense and prolonged radiation, direct and indirect.

(7) Abundant and prolonged light.

(8) Violent winds driving clouds of dust and shifting sands, or, on the contrary, clean and transparent air.

(9) Clear sky and general absence of clouds.

(10) Limited dew formation.

(11) Limited sub-soil water resources.

(12) Soils which are sandy, clayey, salty and rocky; and, possibly, sand dunes and naked and rugged hills and mountains.
(13) Soil temperatures much in excess of air temperatures.
(14) Rarity of vegetation and animal life.

MACROCLIMATE

Complex elements of nature constitute the desert climate. These elements are interrelated, and their influences on biota are exerted in various ways. Climatic factors of stress are directly and indirectly associated with the biological problems of the desert.

According to Duhot (1945), 'climate is the atmospheric conditions observed over a definite region during a long period'. It is estimated that a 35-year period of observation is necessary to study the climate of a locality. As climatology is still in its formative phase, essential data for the explanation of climatic changes are lacking.

Miller (1953) made a distinction between the elements of climate, such as air temperature and humidity, duration of sunshine, rainfall and wind velocity, and the factors of climate (determining causes), such as latitude, altitude, wind direction, distance from the sea, relief, soil type and vegetation. Climatic elements vary, whereas climatic factors are relatively constant.

The phytogeographic and ecological definition of desert climate was given by Emberger (1938) as follows: 'Desert climate is characterized by rains without seasonal rhythm and by long periods (exceeding at least one year) without rainfall.'

The principal elements of desert climate are discussed below.

Aridity

Dryness is the dominant characteristic of deserts. Miller (1953) defined desert climate on the basis of aridity combined with temperature range, according to the formula

$$R(\text{in.}) \leq \frac{T° F}{5}$$

namely, rainfall in inches is less than one fifth of temperature in degrees Fahrenheit. Hence he subdivided desert climates as follows:
(1) Hot deserts, with no cold season; i.e. no months below 43° F (6·1° C).
(2) Cold deserts, with cold season; i.e. one or more months below 43° F (6·1° C).

Solar radiation

Deserts receive abundant and prolonged radiation from the sun. Kachkarov and Korovine (1942) observed that light is intense in desert regions and rich in radiations belonging to the right side of the spectrum (blue, violet, ultra-violet). Also, Barnett (1954) reported that the Sahara shows a cloud cover of only 10 per cent during the winter and less than 4 per cent from June to October.

According to Biel (1944), the dryness of desert air and the cloudless skies cause strong insolation during the day and strong radiation during the night. Even desert stones, unable to withstand the daily expansion and contraction, are broken.

The insolation of the North African littoral is known to exceed 3000 hours a year, and in regions south of the Atlas Range it may average 4000 hours a year. Betier (1958) reported that in Tamanrasset the total radiation received on a horizontal surface averages 540 cal/cm^2 a day.

Air temperature

The few records that are available at present do not give a true picture of the daily, seasonal and yearly regional variations of desert temperature.

In describing the heat of the Sahara, Lhote (1937) said that this desert is an oven during the day and an ice-house during the night. During summer, 52° C in the shade and 70° C in the sun were recorded, but at night the temperature falls to 10° C. In winter, the diurnal temperature is about 22° C and the nocturnal one approximates 0° C.

Barnett (1954) reported that the highest temperatures ever officially recorded were taken in the Sahara at Azizia, Libya (136° F, or 57·7° C, in the shade) and at Death Valley, California (134° F, or 56·6° C).

According to Buxton (1955), the extremes of temperature observed in one year showed an absolute range of 57° C at Ghardaia, Algeria, and 88° C at Kasalinsk, Turkestan.

The daily range of temperature reported by Tanon and Neveu (1934) for the Sahara is 50°C in summer and 17–20° C in winter.

Soil temperature

Desert soil, because of its heat conductivity and heat capacity, is easy to warm during the day and quick to cool during the night.

Kachkarov and Korovine (1942) found, in the Kara Kum desert in May 1925 at 1445 h, a variation of temperature from 49° C on the sand surface to 22° C at a depth of 90 cm. Also, in August 1933, the same authors observed a desert surface temperature at Betpak-Dala of 23° C at 0700 h and of 52·2° C at 1300 h. Ranges of soil temperature from 0° C to 79° C have been observed.

Winds

Among the moulding forces of the desert environment are the winds. They intensify evaporation and dryness, besides transferring masses of dust and sand particles. Their effects on desert plants and animals have not been adequately investigated.

Gautier (1950) reported that the burning wind of the desert, charged with sand and dust, exercises a depressing influence on man and animals. The khamsins of Egypt, the siroccos of Algeria and the Arabian simooms are all local varieties of the same desert wind. Apparently, these winds originate in the deserts and semi-deserts of North Africa, Palestine and Syria, and blow south-east and east.

In his description of real siroccos, Biel (1944) mentioned that humidities as low as 8 per cent have been recorded. Dust is carried extremely high in the air, up to 4 km and more. The hot desert soil temperature often reaches 70–90° C. The atmosphere, full of dust, affects visibility, giving the impression of a dense fog.

The rising sun heats desert areas quickly, giving rise to warm air currents (the specific heat of land being 0·6 of that of water, for every 6° C rise in water temperature land temperature rises 10° C). Winds increase with the intensity of radiation, thus creating dust storms. The estimated average wind velocities are between 10 and 15 miles per hour, but frequently 20–30 miles per hour have been recorded.

Humidity

Inland deserts have an extremely low atmospheric relative humidity compared with that of coastal deserts. This important factor affects both plants and animals. Also, the daily range of relative humidity in the desert is greater than the annual range. This variation is caused by complex and interdependent factors, such as intense heat, excessive and prolonged radiation, quick evaporation, winds, scarcity of rain and vegetation and distance from water masses. The problem of atmospheric humidity is aggravated when the combined effects of high air temperature and low relative humidity exercise additional strain on the water balance of plants and animals.

According to Biel (1944), the coastal region of Oran, Algeria, in July shows a monthly average maximum temperature of 34·4° C and a relative humidity of 74 per cent, whereas the inland region of Biskra during the same time shows 45·6° C and 33 per cent. Gardinier (1958) reported that the average relative humidity in the Sahara at Hoggar is 23 per cent at 1800 h and on certain days it may even drop to almost zero.

The atmospheric vapour pressure of the average desert region ranges between 10 and 15 mm Hg. According to Siple (1949), it may, in the driest deserts, drop to below 1 mm Hg. This low humidity permits rapid evaporation, which is further accelerated by air movement. At present no records are available to show to what extent temperature rises and atmospheric humidity drops in the heart of the Sahara.

Evaporation

The rate of evaporation in deserts is rapid because of the prevalence of high temperatures, intense radiation, low atmospheric humidity, cloudless skies, continuous breeze or winds and bare soil surfaces.

The ratio of evaporation to rainfall (e/r) is also high. Kachkarov and Korovine (1942) reported that at Ghardaia, Algeria, the ratio of evaporation to rainfall is 59·7 (annual evaporation 5309·0 mm, rainfall 88·7 mm). In the Libyan desert evaporation reaches 4000 mm a year, while rainfall takes place only once every four or five years.

Evaporation rate varies with situation and locality. This is important from the bioclimatic point of view, as desert animals seek shelter in burrows and caves. Williams (1923), who investigated bioclimatic conditions in the Egyptian Desert at Wadi Digla, near Cairo, reported the following data for the average daily evaporation in August (using Piche evaporimeters): on the plateau 21 mm, in an open wadi 17 mm, in the shade of the camp 12 mm and 12 m inside a cave 2·9 mm.

Dew formation

The fact that a limited amount of dew is formed in the desert may be attributed to the low vapour content of desert air, the rapidity of evaporation, the rarity of rains and vegetation, and various other factors. Nevertheless, the little dew that is deposited may play an important role as a source of water for plants and animals.

Finbert (1938) expressed the view that during the years of dryness in the desert the plants depend on dew for their existence. The fog moistens the soil, and the sand has the power of retaining the humidity around the roots of plants.

Recently, Dr F. W. Went, of the Plant Research Laboratory of the California Institute of Technology, has reported that various plants, such as tomatoes, sugarbeet, peas, squash and mint, are able to absorb water from dew at night and store it in the soil for later use by exuding it through their root system.

Rains

There are very few systematically recorded data on rainfall in desert areas. In general, the annual rainfall on deserts is estimated to average 100 mm. However, these rains are irregular and several years may pass without any rainfall. There are also desert regions that are entirely devoid of rain and where no life can exist.

Tanon and Neveu (1934) reported that the annual rainfall in the Sahara is between 15 and 20 mm, but in some parts it rains only once in five or nine years. Capot-Rey (1953) has reported the following annual rainfalls: Tamanrasset, 40·7 mm; Tindouf, 32 mm; Bilma, 17·6 mm; Adrar, 14·8 mm.

Desert soils and moisture

Among other factors, the processes of erosion and corrosion give rise to different types of desert soil. The diurnal, nocturnal and seasonal heat variations, the violent winds and the occasional torrential rains serve as perpetual forces, levelling elevations and transferring sand and clay.

From the point of view of vegetation, what is said to be lacking in desert soils is humus and proper drainage: different types of desert soil have their limitations for the sustenance of plant life. They differ in texture, porosity, capillarity, moisture content, chemical content, their proximity to water supplies other than rain and dew, and the degree of their exposure to the sun and winds. All these considerations, and many others, contribute to the development of specialized types of desert vegetation.

Killian and Feher (1938) have found microorganisms (bacteria, champignons, algae) that are adapted to the dryness and the extreme temperatures of the Sahara desert soil, vegetating even during the seasons that are most unfavourable to life.

MICROCLIMATES AND DESERT LIFE

What interests the biologist is not only the macroclimate, as discussed above, but also the infinitely more important microclimate, close to the ground and underground, which plants and smaller animals, individually and in communities, adopted as their desert habitat to attenuate the severity of macroclimatic influence.

Prenant (1934) wrote: 'The fundamental idea of ecology is that of adaptation; namely, of certain correlations between the organism and its milieu. The essential fact of adaptation seems to be the prodigious capacity of the living matter for expansion.' As Monod (1953) indicated, the object of biological investigation is to define the common relationships of biocenoses, biotopes, microclimates, nature and morphology of the substratum, etc.

Desert biota

Biogeographic studies describe the distribution of biota on earth. It seems that each species has its preferred habitat. In natural succession, tundra, temperate forest, steppe, desert, savannah, tropical forest and equatorial forest spread from pole to equator.

As deserts are poor places for the preservation of fossils, paleontological evidence on the history and development of biota is scarce. Moreover, deserts have appeared and disappeared since Permian time, and little is known of how existing desert areas were related in the past.

According to Shreve (1934), each of the great deserts has its own biological history. The deserts were populated by species of plants and animals from the neighbourhood rather than from other deserts. These species developed various methods of adaptation to the arid conditions. Shreve draws the following taxonomic conclusions:

'(a) The desert has fewer species of plants than the moist regions, about the same number of species of birds and animals and a greater number of species of reptiles. (b) The organisms of the desert show varying degrees of relationship to those of moist regions. (c) Most of the species, many of the genera and a few of the families are distinctive. (d) These degrees of relationship are unlike in America, Asia, Africa and Australia.'

Because of its isolation from the rest of the world and its special climatic elements, the desert is considered to be one of the principal biotic regions of the earth. Its biocenosis includes distinctive vegetation and animal life that have undergone specialized adaptation to the rigorous conditions of the desert environment.

According to Kachkarov and Korovine (1942), the biocenosis of the desert consists of a nucleus of vivacious plants and animals that are well adapted; namely, hoofed animals, rodents, certain carnivora, lizards and serpents and, among birds, the *Podoces panderi*. Seasonal forms, such as the migrating birds, come to build their nests around this sedentary nucleus, and numerous migrating birds traverse the desert in the spring and in the autumn.

Nowhere is the dependence of animal life on vegetation more evident than in the deserts, for plants provide the animals with their essential food, water and shelter. The battle against drought is first won by the plants, and where plants exist there are animals. The transition from a semi-desert to desert environment was described by Cailleux (1953) as follows: 'As dryness is accentuated, plants become rarer and, by transition, semi-desert and desert lands are reached where flora and fauna are very poor. Plants are tough and thorny, or, on the contrary, provided with reserves of water. Animals are nocturnal. Their colour resembles that of the soil. Their enlarged feet facilitate walking on the sand.'

Plants and animals have developed various methods of adaptation to their desert environment. The advantage that animals have over plants is their mobility. Plants are fixed to the ground and have no choice but to adjust themselves to their immediate environment, or die. But the mobility of an animal enables it to look for a more favourable environment.

The important fact to note here is that both plants and animals have managed to resist the hostile elements of the desert climate by developing various mechanisms of adaptation. These adaptations may be broadly classified as ecological, behavioural, functional, structural and genetic.

Desert flora

Rains, irrigation, or both, transform desert areas into fertile lands. What limits vegetation in hot deserts, therefore, is dryness. Given an adequate humidity in the soil and air, the other factors of heat, light and minerals favourable to plant development are more or less available. Hence during the dry season and the period of drought, only those plants which have developed adaptive mechanisms to dryness will survive. It is estimated that not more than one fifth of the vast desert flats is covered with vegetation.

According to Buxton (1955), who studied desert flora in relation to animals, the three most important groups of plants are as follows:

(1) The annuals, or ephemeral, vegetation, whose seeds survive a dry period of months or years and whose stems, leaves and flowers are not structurally modified to resist dryness. They appear after rainfall and are short-lived. Where two rainy seasons exist, in winter and summer, the seeds of winter annuals germinate in winter while the seeds of summer annuals lie dormant (hibernate), and conversely, the seeds of summer annuals germinate in summer while the seeds of winter annuals lie dormant (estivate).

(2) Plants which grow leaves and stems and flower after rainfall. The lower parts exist throughout the year beneath the ground as a bulb, corm, tuber or fleshy root for storing food.

(3) Plants which have made adequate structural adjustments to exist above ground during all seasons. These have developed specialized structural changes in order to reduce loss of water from the surface or store it in the body of the plant, or both. Gautier (1950) wrote: 'All Sahara plants have the ingenuity of defending themselves against dryness. Lying low, protected from winds, deprived of leaves or equipped with small spiny leaves, chlorophyll massed in their fleshy twigs, the small reservoirs of fluid, they have developed roots of unbelievable size in order to go deep in search of water.'

In his description of the plant life of the Sonoran Desert (Central America), Shreve (1936) reported three main types of plant: (1) the ephemeral plants, which depend upon rains during the two rainy seasons, at midsummer and late in winter, for their different temperatures of germination; (2) the succulent plants, which have developed special structural features for water economy and water storage (among these plants are the cacti, with more than 1200 species); (3) the non-succulent plants, which have a larger number of species and greater variety of structure. Their problem of water balance is acute.

We may conclude from these findings that the desert flora have found various ways of adjusting their vital processes to the severe conditions of the environment, and that in the course of their long struggle, as species and as communities, they have developed specialized features distinctive of the desert biotic region to which they belong.

Desert fauna

For the study of desert animal life the two principal factors that need consideration are the periodical climatic changes and the ability of animals to move from place to place. Furon (1951) has demonstrated that the instability of the quaternary climate induced variations in fauna. The rainy periods of Africa correspond to the glacial periods of Europe. There exist three faunae: one of cold countries, the second of steppes and the third of hot countries; but there exists also a mixture of all these kinds, according to the capacity of the animals to adapt.

The prehistoric fauna of large animals, which inhabited many parts of North Africa until the neolithic epoch (elephant, rhinoceros, giraffe, large antelopes, etc.), is now rare and found only in isolated places. More arid conditions have succeeded this prehistoric period, and with them the distribution of animal life has changed.

The actual fauna of the Sahara, wrote Lhote (1937), is poor in species. Among other animals, there exist antelopes, gazelles, small carnivora, jackals and foxes, jerboas, small rodents, reptiles, birds and small insects. Many of the animals are runners, jumpers or burrowers. Also, their hair is of a light colour resembling the colour of the desert soil.

Aridity being the dominant factor in the desert, the principal problem of animals is the acquisition of sufficient water for their metabolism. According to Dekyser and Derivot (1959), among the land animals of the Sahara there are those which drink regularly, those which drink irregularly, those which are satisfied with the preformed water found in their food, and those which do not drink.

The rhythm of desert life is to a great extent dictated by the daily and seasonal cycles of the sun and the irregular cycles of rainfall. The severity of climatic stress on the vital processes of organisms is closely associated with these cycles. This is why desert animals everywhere have found generally similar ways of solving their problems of adjustment.

Most desert animals are nocturnal; they pass the day inactively, hiding in their places of shelter or sleeping in their own or in other animals' burrows. They find protection in the shade of every possible refuge, such as stones, niches, holes, crevices, caves and bushes. Desert animals which spend the day dormant are active at night in search of food and prey. The carnivora devour the herbivora, the mighty and crafty prey on the small and weak. It is the small world of rodents that provides the main food of birds, reptiles and mammals. The herbivora try to satisfy their hunger and quench their thirst by whatever vegetation and grain they are able to find.

MORPHOLOGICAL AND PHYSIOLOGICAL ADAPTATIONS

Coloration

On the much-discussed question of the colour of desert animals, Buxton (1955) expressed the view that the prevailing buff, sandy or pale grey colour is the result of the conditions of desert life.

However, Heim De Balsac (1936) observed similarity of coloration in only 24 out of 47 species of bird and 39 out of 50 species of mammal. As our present knowledge on the subject is limited, it cannot be stated precisely whether it is climate, nutrition, protective mechanism or some other factor that determines animal coloration.

Morphological character

Numerous references exist in the literature to the morphological characters of desert animals, which certain authors attribute to the influences of the environment. The following are some examples: fringed toes for locomotion on loose sand; valve-like closure of nostrils, eyes and mouth for burrowing in loose sand; reduced number of toes; elongated hind-legs; widening of the bodies of lizards and snakes for lateral and vertical movement; inflated bullae tympani; relatively impervious skin; highly developed organs of hearing, vision, smell and touch; texture of hair or fur.

Arguments for and against these supposed adaptations have long been discussed but their causes remain to be discovered.

Dormancy

The phenomena of hibernation and estivation among desert animals are known. Kachkarov and Korovine (1942) have reported that in 1929 Kalaboukhov verified experimentally that susliks fed on dry food lapsed into lethargy; namely, their respiratory movement fell from 60–80 to 2–5 per minute and their body temperature remained about 2° C above the air temperature, between 12° C and 22° C. These authors also mentioned that reptiles and insects pass through a hibernal torpor, and that during the period of dryness many desert animals become dormant after storing a good reserve of fat in their body. The water of metabolism suffices for their needs.

Water needs and heat tolerance

Evidence is rapidly accumulating on the ability of many desert animals to live without drinking water. In addition to carnivora, which receive their supply of water by devouring warm-blooded vertebrates, there are insect-eating birds, lizards and serpents that are contented with the water that they derive from their prey.

Mason (1936), in an account of the expedition that he and his party made by motorcar in 1935 across 6300 miles of the Libyan Desert, mentioned that Wadi Hawar (Sudan), which is the relic of an ancient river, is really the Libya of Herodotus, populated by wild animals, including addax, adda gazelles, dorcas gazelles, oryx, lions, mountain sheep, hyenas, ant-bears and porcupines, foxes, jerboas, gerbils, hussar monkeys, ostriches and bustards. He and his colleagues believe that these animals exist lifelong, or at least the greater part of the year, without drinking. Furthermore, Kachkarov and Korovine (1942) indicated that according to Chapman (1921) the antelopes (addax and oryx) and the gazelles (adda and ariel) never drink, and that according to Andrews (1926) the wild monkeys of the Gobi Desert also do not drink.

In their publications on desert animals, K. and B. Schmidt-Nielsen (1952) stated that the small rodents dwelling in the Great Palearctic deserts (*Dipus, Jaculus, Gerbillus, Meriones, Dipodillus*), in the South African deserts (*Pedetes*), in the American deserts (*Dipodomys, Perognathus*) and in the Australian deserts (*Notomys, Ascopharynx*) have certain common features. Besides bipedal jumping locomotion, long hind-legs and a reduced number of toes, they have the capacity to live on dry grain, without access to drinking water.

The camel

This 'ship of the desert' has probably been the object of more study and serious investigation than any other large desert animal. There are many publications on the life, qualities, anatomy, physiology and pathology of the camel. Among others, the following authors have collected useful information on these subjects: Steel (1890), Cauvet (1925), Robinson (1936), Finbert (1938), Curasson (1947).

According to Lhote (1937), camels can survive without water for six days during summer and for 18 or 20 days during winter, when the environmental temperature is low.

Schmidt-Nielsen and co-workers (1959) made a special investigation of the camel's physiological mechanisms of heat tolerance and water economy. They found that the camel's desert hardiness is due to the following facts:

It is able to tolerate a water loss (dehydration) of 25–30 per cent of its body weight; that is, about 40 per cent of its total body water. The daily range of its body temperature variation is between 93° F (33·9° C) and 105° F (40·5° C), i.e. 6·6° C. Sweating does not take place until the animal's body temperature reaches 40·5° C; hence the camel is able to store much more heat before feeling the need to lose it through evaporation and sweating. The insulating power of its hair and of the fat accumulated on its back reduces its heat gain from the environment. It also economizes water by diminishing its sweat and by excreting concentrated urine and dry feces. It is capable of drinking 27 gallons of water in ten minutes to satisfy its thirst.

The kangaroo rat

The studies of Schmidt-Nielsen (1948–1958) on the kangaroo rat, *Dipodomys*, of the New World deserts, provided valuable information about the water conservation of small desert rodents.

Dipodomys, when deprived of water and fed exclusively on dry barley, can maintain its weight and water balance through the following physiological mechanisms:

(1) The total evaporation (cutaneous and pulmonary) of *Dipodomys* is 0·54 mg of water per millilitre of oxygen consumed, whereas that of the white rat is 0·94 mg. This water economy arises from the lesser cutaneous evaporation of *Dipodomys* compared with that of the rat.

(2) The urine of *Dipodomys* is more concentrated than that of the rat:

	Urea	Electrolytes
Dipodomys	23%	7·0%
Rat	15%	3·5%

(3) On the metabolization of 25 g of dry barley—namely, 100 kcal—the feces of *Dipodomys* contain 0·76 g of water, whereas those of the rat 3·4 g.

(4) *Dipodomys* maintains its water balance through the preformed water and the water of oxidation in its food, whereas the rat cannot live without drinking.

(5) *Dipodomys* economizes water by spending most of its time in burrows. where the air is relatively more humid than the air outside.

The jerboa

Our own studies on the thermoregulation of the jerboa, *Dipus aegyptius*, a rodent of the Egyptian western desert, compared with the white rat, a rodent of temperate regions, have given the following general results:

(1) The jerboa has adopted an underground habitat and a nocturnal way of life in the desert.

(2) It is capable of living from one to three years on dry diet, composed of dry grains of wheat and barley (without water), whereas the white rat does not tolerate the dry diet more than three days.

(3) Its body temperature is lower than that of the rat by 0·75° C (jerboa, 36·8° C; rat, 37·55° C).

(4) Its basal metabolism is 3·649 kcal/kg/h compared with the rat's 6·156 kcal/kg/h—a difference of 2·507, or 68 per cent.

(5) Its evaporation at the environmental temperature of 30° C is 0·659 g/kg/h compared with the rat's 1·136 g/kg/h—a difference of 0·477, or 72 per cent.

(6) The jerboa easily tolerates environmental temperatures up to 45° C by entering into a state of lethargy (deep sleep) beginning at 35° C, whereas the rat's temperature tolerance limit is 40° C.

(7) At the environmental temperature of 40° C, the jerboa's heat production reaches 5·595 kcal/kg/h and its evaporation 8·268 g/kg/h, whereas the rat's reaches 10·892 kcal/kg/h and 14·742 g/kg/h.

(8) At the environmental temperature of 25° C, the jerboa's and rat's 24-hour heat balance and water balance are as follows:

(body weight 150g)	Heat balance (kcal/24 h)	Water balance (g/24 h)
Jerboa on dry diet	20·9	3·6
Jerboa on wet diet	33·5	10·6
Rat on wet diet	38·9	17·9

CONCLUSIONS

(9) The jerboa accepts both dry and wet diets, eats little and excretes little. Its urine is concentrated and its feces are dry. The rat accepts only wet diet, eats much and excretes a great deal. Its urine is watery and its feces are moist.

CONCLUSIONS

The above-mentioned findings on the general ecology of the desert indicate:

(1) that the severity of desert conditions exercised varying degrees of stress on the vital processes of plants and animals;

(2) that a combination of rigorous environmental factors created complex problems of existence for desert biota;

(3) that in their struggle for survival, whether as individual species or as communities, they have found different solutions to desert problems and developed, in the course of evolution, various mechanisms of adaptation;

(4) that, in order of prevalence, their adaptations can be characterized as ethological, ecological, physiological, morphological and genetic.

The chapters that follow present the findings of our comparative study on the jerboa and white rat.

2
JERBOA AS A DESERT SPECIES

ORIGIN

The desert animal that we have studied is *Dipus aegyptius* or *Jaculus orientalis* (Jerboa), a species of the genus *Jaculus* from the family of *Jaculidae* or *Dipodidae*. The family of *Jaculidae* belong to the suborder *Simplicidentata* of the order of *Rodentia*.

According to Brehm (1890), the Egyptian jerboas were known to Ancient Greeks and Romans as bipedal rats. Anderson (1902) reported that Herodotus, Theophrastus and Aristotle referred to this desert rodent in their writings, and that it appears in the literature under various names, as follows: *Jaculus orientalis*, Erxleben (1777); *Dipus jaculus*, Gmel. Linn. (1788); jerboa, Bruce (1790); *D. sagitta*, Schreb (1792); *D. Abyssinicus*, Meyer (1793); *D. gerboa*, Olivier (1801); *D. bipes*, Licht. (1823); *D. aegyptius*, Licht. (1825); *D. (Haltomys) oegyptius*, Brandt (1844).

The biological history of *D. aegyptius* raises the following questions: What is the origin and evolution of the existing species of jerboa? When did they first appear in Africa? Which species of the family *Dipodidae* is the most primitive? There are no clear answers to these questions, but certain indications exist which may provide probable explanations.

Romer (1954) divided the Age of Mammals, or the Cenozoic Era, into seven epochs, extending from the Paleocene (55 million years) to the Quaternary Recent (20,000 years). The same author stated that rodents are the most successful of all living mammals. The extent of their distribution and adaptation is the widest. They have no aquatic forms. Their bodily structure is little specialized. The functions of gnawing and chewing characterize their group.

About the branching out of rodents from the early placental mammals in the course of evolution we have no information. We know, however, that the Cenozoic Era, or the Age of Mammals, includes two periods of geologic history; namely, the Tertiary Period, which began 55 million years ago, and the Quaternary Period, which began one million years ago. On the basis of certain paleontological evidence it seems probable that the origin of rodents dates back to the middle of the Eocene Epoch, about 35 million years ago.

According to H. and G. Termier (1952), the Oligocene grottoes, which were filled with 'phosphorites of Quercy', included, among other mammals, small rodents. Many forms of life were found, such as *Paramys*, of the Eocene, arboreal squirrels, marmots, jerboas and rats.

The problem of the centres of distribution of fauna is a complicated one. The subject of intercontinental migration lends itself to much speculation. Guiart (1934) advanced the following view: As South America remained united to Africa until the end of the Tertiary Period, a migration of edentata and hystrichomorphic rodents took place. Also, as land connections existed between Europe and Africa at the beginning of the Quaternary Period, Europe was able to receive its fauna from Asia and Africa.

DISTRIBUTION

In his classification of the families and genera of living rodents, Ellerman (1949) places the family of *Dipodidae* as follows: *Dipus*, Pleistocene (Asia); sub-family *Dipodinae*, Lower Pliocene–Recent (Asia), Pleistocene–Recent (Europe) and Recent (Africa).

According to Furon (1950), during the glacial periods jerboas inhabiting the steppes of Asia and Russia advanced as far as the neighbourhood of the Rhine Valley.

Meinertzhagen (1930), who summarized the principal climatic and geographical changes in Egypt from the beginning of the Quaternary Glacial Period, wrote that about 300,000 years ago desert conditions prevailed throughout Egypt. The fauna and flora of the oases owe their origin to the climatic periods which preceded the advent of desert conditions; 90 per cent of the desert mammals are species not found elsewhere. Egypt's principal elements are Palearctic, but tropical and African forms enter at Aswan and Wadi Halfa (Sudan). In Sinai, Asiatic forms are found. From the west, through Alexandria, Cyrenaic forms infiltrate, and the Sahara type of fauna and flora reach the Nile Valley. Along the coasts, mediterranean forms are found.

The foregoing considerations seem to indicate that an Asiatic form of rodent, closely resembling *Dipus aegyptius*, invaded North Africa sometime in the distant past and that, as desert conditions supervened, there gradually evolved the present desert species which we identify as *D. aegyptius* or *Jaculus orientalis*.

DISTRIBUTION

According to Sanderson there are 12 genera of *Dipodidae*, and a large number of different species exist in Eastern Europe, Russia, Central Asia and North Africa.

In his classification of Egyptian rodents, Innes (1932) mentioned that of the 13 families distributed all over the world, seven are represented among the fauna of Egypt, i.e. *Muridae, Jaculidae, Spalacidae, Sciuridae, Myoxidae, Hystricidae* and *Leporidae*.

The two prevalent species of jerboa in Egypt, which have a close mutual resemblance and which exist in two different areas, are *Jaculus jaculus* (Linnaeus, 1758) and *Dipus aegyptius* or *Jaculus orientalis* (Erxleben, 1777). Ellerman (1949) said that the latter is Palearctic (North African) in its distribution.

In his description of the Libyan Desert, Bagnold (1954) mentioned that the jerboa is found in places where no other form of life exists.

According to Wassef (1953), *J. jaculus* is widely distributed in all the deserts and oases of Egypt down to Sudan and Somaliland, to the east and west of the Nile River, over North Africa and to south-west Asia. It is found in Iraq, Arabia, Palestine, Syria, Egypt, Libya, Sudan, Tunis, Algeria and Morocco. The *D. aegyptius*, however, has a limited distribution west of Alexandria along the littoral belt. It is found in Egypt, Tripoli, Tunis and Algeria.

There is another genus of the family *Dipodidae*: the *Scirtomys tetradactylus* (Lichtenstein), or *Allactaga tetradactyla*. This jerboa has four toes, while both the *J. jaculus* and *D. aegyptius* are three-toed rodents. It also has longer ears and a smaller tail than the other two jerboas, and its size is intermediate between the two.

Wassef (1953) reported that according to De Beaux (1931) *A. tetradactyla* is known only in the regions of Maryut and Matruh and in El Agheila, and that Mersa Matruh is the only known locality in Egypt where all three species of jerboa exist.

DESCRIPTION OF THE JERBOA

Of the three species mentioned above, in the order of body size, *Dipus aegyptius* is the largest, then *Scirtomys tetradactylus*, and *Jaculus jaculus* is the smallest.

The tail of *D. aegyptius* is longer, its snout broader, its cheeks more prominent and its ears longer and broader than those of *J. jaculus*. Moreover, the remarkable feature in the anatomy of *D. aegyptius* is the two horny spines, about 9 mm long and 1 mm thick, attached to the dorsum of its glans penis. This peculiar armature is completely lacking from the penis of *J. jaculus*. The mammalogist Dr H. Hoogstraal is of the opinion that the three-toed *D. aegyptius* and *J. jaculus* are well adapted to the desert, and that the four-toed *Allactaga tetradactyla* is less adapted and more primitive.

In order to gain an impression of the comparative sizes of these three species of Egyptian jerboa, we have prepared *Table 2* from the figures quoted by Innes (1932).

Table 2

Sizes of Egyptian jerboas (in millimetres)

	Dipus aegyptius	Jaculus jaculus	Scirtomys tetradactylus
Snout to root of tail	150	110	118
Length of tail	230	180	150
Length of head	41	35	—
Height of ear	28	18	33
Width of ear	—	15	15
Length of hind foot (without claws)	77	60	57 (tarsal)
Length of crane maxima	40	34	—
Width of crane maxima	30	24	—

It seems, therefore, that the smallest and the most widely distributed of the three jerboas, the *J. jaculus*, is probably an African type adapted to very dry and hot desert climate; that the largest jerboa, *D. aegyptius*, is probably an Asiatic type, adapted to the more humid desert climate along the littoral belt of North Africa west of Alexandria; and that the intermediate-size jerboa, the *S. tetradactylus*, is the most ancient type but not as successful in its adaptation to desert conditions as the other two.

Hatt (1932) has studied the vertebral columns of more than 60 bipedal leaping rodents, among which were the Egyptian jerboas mentioned above.

These rodents belonged to 18 genera. He reached the conclusions that the vertebral anatomy of jumping rodents reflects remote genetic background; that the equ'ibrium of the bipedal animal is maintained by caudal displacement of the centre of gravity, which is occasioned by the shortening of the neck, the increase in cervical flexure, the reduction of the forelimbs and the increase of tail weight; and that the tail serves for support when the animal is standing upright, for lateral control of balance during progression and for counterpoise.

According to Buxton (1955), in each continent animals are found which show a reduction in the number of toes. In the case of *J. jaculus*, the reduction has gone so far as to leave only three toes, without any vestige of the other two. These genera are found in the great Palearctic Desert and the neighbouring steppes.

Anatomical features (Figure 1)

Head: The jerboa's forehead is convex. Its eyes are large and placed laterally near the ears. The ears are broad and about 2·8 cm long. Its snout is short, broadening toward the eyes. Its nostrils are surrounded by thickened skin, and the structures around the orbital cavities protect the eyes when the head is used, in conjunction with the hands, for shovelling soil from the burrow or beating it to solidify the burrow structure. The length of the head is about 4·1 cm.

Neck: The fusion of cervical vertebrae gives the jerboa a short neck, which, according to Hatt (1932), helps in body balance.

Trunk: The jerboa's chest is funnel-shaped, and its abdominal cavity is narrow and short. The length of the trunk is about 15 cm.

Hands: The jerboa's forelimbs are short and bent close to the head. The arm and forearm are very short. The hand is small with five thin fingers, ending in long, curved and sharp claws. The palm is nude and concave in shape, with two tuberous protrusions toward the wrist. The back of the palms and the sides of the fingers have white hair. The forelimbs are used as hands and play no part in locomotion.

Hind-limbs: The jerboa's hind-limbs are about four times as long as the forelimbs. They are specially adapted to quick jumps. The relative length of the hind-limb bony structure and the fact that the metatarsal bones are fused together provide powerful leverage to muscular action. According to Alezais (1900), the muscular insertion of sprinting animals like the jerboa is concentrated near the proximal extremity of bones.

Each foot has three laterally flattened long toes, ending in short but strong and somewhat convex claws. The sides of the toes and particularly the bottoms are covered with brush-like hairs, which help locomotion on sand and the backward displacement of soil in burrow digging.

The hind-limbs are provided with long and strong muscles. They enable the jerboa's speedy locomotion and long sprints.

Tail: The long and graceful tail of the jerboa ends in a black and white haired tuft. The length of the tuft alone is about 8 cm.

The tail structure is quadrangular at the root, becoming round and thin towards the tip. It is covered with short and thin hair. Its length is about

23 cm. The tail serves for body balance during locomotion, for support when the animal is standing upright and for heat dissipation.

Fur: The jerboa's body is covered entirely with thin and silky hair. The ventral parts are almost white and the dorsal parts, including the head,

Figure 1. The jerboa (*Dipus aegyptius* or *Jaculus orientalis*).

are yellowish-brown. Black, yellow and grey tips scattered irregularly in the dorsal parts add variety to its colouring. Both the inner and outer parts of the ears are covered with thin and short hair. The eyelashes and the long sensory hairs are black. The whiskers are a greyish-white.

Morphological observations

The jerboa's body measurements and features mentioned in the previous section must be supplemented by the following observations.

Body weight: In general, the body of the male jerboa is larger and heavier than that of the female. Appendix 2 indicates that among the animals kept in the laboratory the average weight of 25 male jerboas was 166·1 g, whereas that of 25 female jerboas was 160 g. While the extreme values are of the same order—namely, 144–190 g for males and 141–190 g for females—it is noteworthy that 28 per cent of the males and only 16 per cent of the females have their body weights included in the 180–190 g bracket, whereas 40 per cent of the females and only 12 per cent of the males have their body weights included in the 140–150 g bracket. Between these two weight-groups, the distribution of the extreme weights on the two Gaussian curves is close.

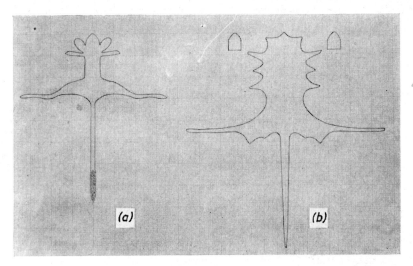

Figure 2. The jerboa's contours: (a) of the body; (b) of the skin area.

Body parts: The body parts of the male jerboa differ slightly from those of the female. These differences are not so much of size as they are of contour. The female jerboa has more elegant lines than the male. Appendix 3 gives some measurements of the body parts of a jerboa, made in a post-mortem study since it is impossible to take measurements on a living animal.

Body contour and skin area: In order to gain an impression of body size, the corpse of a male jerboa was spread on 1-mm graph paper with the ventral side up, and the contour of the largest extension of its members was marked (*Figure 2*). Our measurements were as follows: maximum body length, from nose to tip of tail, 40 cm; width of largest lateral extension of hind-limbs, 35 cm; length of maximum vertical extension of hind-limbs, from nose to extremities of the limbs, 30 cm. These findings indicate the marked capacity of the jerboa to reach out in case of necessity (*Figure 2a*).

To study the skin area of this specimen, the following technique is used: The body is placed on the operation table, with its ventral side up, and three incisions are made: (a) a vertical incision of the skin from the tip of the tail to the snout; (b) a sagittal incision from the tip of one hind-limb to the tip of the other hind-limb, across the abdomen; and (c) sagittal incision from one forelimb to the other, across the upper chest. After these incisions are made, the entire skin can be detached easily, like a mantle, from the animal's body. The skin, once removed, is spread, fur up, on the 1-mm graph paper and its contour is marked. The ears are detached and marked separately. *Figure 2b* shows the skin contour of the specimen.

Using the same technique, we have measured the body surface of adult jerboas whose weights varied from 135 to 189 g. *Table 3* indicates that the ratio of surface (cm^2) to weight (g) is close to 2, varying from 1·9 to 2·13 with a standard error of 0·011.

Table 3

Surface/weight ratios of adult jerboas

Specimen	Body weight (g)	Cutaneous surface area (cm^2)	Surface/weight ratio
J38	135	263	1·95
J16	145	310	2·13
J54	150	302	2·01
J50	153	310	2·02
J12	168	336	2·00
J34	189	375	1·98

Fur: The texture of the jerboa's hair is very fine and silky. The average weight of its total hair is 3·03 g per 150 g of body weight—namely, 2 g of hair per 100 g of body weight.

Mammary glands: The female jerboa has four pairs of mammary glands, situated as follows: one pair at the upper part of the thorax; one pair 3 cm below this, at the level of the tip of the sternum; one pair 4 cm lower; and one pair 3 cm lower, about 2 cm from the vagina.

3
THE JERBOA'S HABITAT

For our study of the desert environment we used the following three methods: visits to desert regions for direct investigation; interviews with the bedouin, desert dwellers and other persons known for their expert knowledge of desert life; and a study of bibliography on the subject.

The instruments used for exploring the jerboa's burrow were: hoe, mattock, shovel, stick 1 m long and two pieces of cotton net about 1·5 m^2; steel measuring tape 2 m long and a 10-m cloth tape; thermometer (maximum and minimum), 'Sydney' dial thermohygrometer, whirling hygrometer with dry and wet bulbs and thermometers for measuring sun and soil temperatures; small tent and cages for animals.

The investigation of the burrows was conducted with the guidance and assistance of the bedouin.

DESERT ENVIRONMENT

The jerboas we studied were captured in their burrows at Maryut, Borg El Arab, El Gharbanyat and El Alamein—the districts of the Egyptian Western Desert which lie along the mediterranean littoral, 30–110 km west of Alexandria. We have personally explored these regions and examined the burrows with the help of the bedouin.

Physiography

The mediterranean coastal zone west of Alexandria has semi-desert physiographic features. There are two lines of limestone ridges, about 2–3 km apart, and extending for long distances parallel to the coast west of Alexandria. According to Ball (1939) these ridges are often as much as 20 m high; they are composed of soft oolitic limestone.

It is south of these ridges, inland, that semi-fertile and semi-desert lands merge into real desert regions, extending far and wide to form the great Libyan Desert plateau.

According to Belgrave (1923), the semi-desert coastal strip, about 60 km wide near Alexandria, diminishes in width as it extends westward for about 500 km to the borders of Libya, where the cliffs of the Libyan Plateau lie near the sea.

The semi-nomadic bedouin, who inhabit this belt with their flocks of goats, sheep and camels, grow barley crops in certain parts of it during the rainy winter season. South of the belt extend for thousands of kilometres the waterless wastes of the great Libyan Desert. The extreme aridity of the Libyan Desert and the absence of any form of life from the wide expanses of its plateau make it the most arid desert of the world.

Climate

The coastal zone west of Alexandria, because of its proximity to the sea, has a mediterranean type of temperate climate: mild weather, humid air, abundant sunshine. There is a two-seasonal periodicity of short rainy winter and long dry summer.

However, a few miles inland, to the south, as the distance from the sea increases and desert areas approach, rainfall becomes rare, the temperature rises, the air becomes dry and the general characteristics of the Saharan type of climate appear.

Rainfall: With increasing distance from the coastal belt, rainfall decreases and the rainy season is confined to the winter months. While the coastal areas may have 20–60 rainy days, inland, at the height of Cairo, there may be only 5–20 rainy days a year. It has also been observed that the maximum rainfall occurs every 7 or 8 years. This was the case in the period between 1929–1930 and 1937–1938.

According to De Cosson (1935), there are years when the rainfall is very scant. For example, in the Maryut area rainfall varies from 40 to 260 mm. Shafei (1952) reported that the average rainfall on the coast is 145 mm; but periods of drought lasting several years were recorded, for example, every year during 1931–1934, when barley crops failed because of insufficient rainfall.

Winds: The prevailing wind is northerly, except during the spring when heat waves and dust storms occur. The temperature at times exceeds 43° C during the khamsin winds, which may persist 3–7 days. Jarvis (1936) reported that during a hot windy April day in Cairo the temperature reached 43–46° C.

Khamsins (from Arabic *khamsun*, fifty) are strong winds which occur during the 50 days following the spring equinox. These very dry southerly winds may be followed by dust storms which usually last a few hours. According to Biel (1944), about 12 khamsin wind depressions are known to traverse Egypt each year. The severest and hottest ones occur in May and early June.

Temperature and humidity: The coldest month of the year is January, when the minimum mean temperature is about 7° C although it rarely reaches the freezing point. The weather becomes warm in April. Heat waves occur from March to early June.

According to Sutton (1946), the highest temperatures recorded were 45° C on the coast, 46·7° C in Cairo and 51·3° C in Aswan.

In general, humidity—namely, moisture in the air—diminishes as one goes south from the Mediterranean Sea. *Figure 3*, prepared on the basis of the Meteorological Report for 1947, shows the temperature and relative humidity variations in the coastal area of Mersa Matruh compared with the desert area of Almaza, 250 km south. The air is considerably drier and warmer in Almaza, particularly during the month of July. The peak of variation comes at 1400 h, when Almaza shows 35·1° C and 38 per cent and Mersa Matruh 28·7° C and 63 per cent temperature and relative humidity, respectively.

The above-mentioned temperatures and relative humidities differ considerably from those recorded in the deserts. Biel (1944) reported that the daily range of desert temperature is greater than that between summer and winter. The dryness of the air causes strong insolation by day and strong radiation by night. For similar reasons the daily range of relative humidity is greater than the annual range, particularly in summer.

Migahid and Abd El Rahman (1953), using Emberger's formula to express the degree of climatic aridity

$$Q = \frac{R}{(M+m)(M-m)} \times 100$$

where R is the annual rainfall, M the mean maximum temperature for the

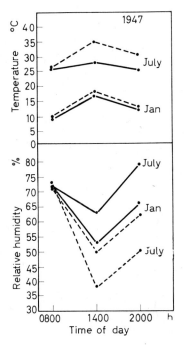

Figure 3. Variations of temperature and relative humidity at the coastal region of Mersa Matruh (———) and the desert region of Almaza (— — — —).

warmest month and m the mean minimum temperature for the coldest month, have calculated the following values for the desert near Cairo:

2·5 in 1948
1·9 in 1949
0·8 in 1950

It is seen that important variations occur from one year to the other. The same authors also made the following observations near Cairo:

Air temperature (summer maximum)	43° C
Soil surface temperature (summer maximum)	63° C
Relative humidity (several hours a day in summer)	20%
Evaporation rate (maximum for July 12–13, 1949)	1·35 mm/h
Air movement rate (January 12–13, 1949)	42 km/h

Rainfall very scanty and limited to 3 months in the year

The severity of these climatic conditions on plant life is evident. Migahid and Abd El Rahman concluded, therefore, that the Egyptian Desert is one of the most arid deserts in the world.

Vegetation

The flora of the desert regions west of Alexandria is of the mediterranean type, being related to Cyrenaic and Algerian forms. Tadros (1953) related that the halophytic communities of Mareotis (west of Alexandria) in their major characteristics resemble species found in other mediterranean countries.

When rainfall is sufficient, a great variety of wild flowers grows in the spring. According to Montasir (1942), the Maryut district is rich in plants. About 800 species have been identified, representing 50 per cent of the Egyptian flora. Meinertzhagen (1930) reported that of the 1514 forms of plant life in Egypt, 668 are found in the desert and 193 of these belong to the desert variety.

THE BURROW AND THE JERBOA'S ECOCLIMATE

The burrow

The comfortable microclimate which the jerboa has created for itself in the midst of the desert, and in which it lives, is its burrow. Anatomically equipped with a pair of hands, each of which has five fingers ending in thin and sharp claws, the jerboa digs fast and deep into the desert sand and clay, using its hands, nose, head and long hind-limbs with highly coordinated movements. The soil so dug is either brushed away with the brush-like hairy toes of its hind-limbs or pushed out with the hands, nose and head, or beaten with the nose and head to solidify the sides of the tunnel. Its flat and muscular nose is a good beating and pressing organ against the soft soil.

The jerboa prepares its burrow with amazing skill and rapidity, operating its hands and hind-limbs with its body in position of maximum extension. Given the environment for burrow-digging, the jerboa can demonstrate all these skills in the laboratory.

Winter and summer burrows: For the rainy winter days the jerboa digs its burrow on the slope of hills to avoid inundation by rain. For the summer it prepares its burrow on less elevated areas, near the edges of open fields, where some vegetation is found.

Burrow types: Among the burrows excavated for the purpose of this study, four main types were discovered, as illustrated in *Figure 4*.

Type 1: Horizontal burrow, which extends about 1·50 m to the interior; its abode is located 1·25 m from the surface (elevation).

Type 2: Angular burrow, which is 2·50 m long and about 1·50 m deep.

Type 3: Zigzag burrow, which extends about 2 m and reaches a depth of about 1·75 m.

Type 4: Hooked burrow, which has a length of 4 m; its abode lies at a depth of 0·75 m from the surface.

The end of each burrow forms a spherical abode of about 15 cm diameter, where the jerboa lives. This nest contains canvas wool in winter months to help the jerboa to keep warm. The wool is prepared by the animal from any piece of jute or cloth that it may find in the fields. Its fine fingers and nails are adept in tearing apart any piece of cloth to make of it the finest heap of wool within a few hours.

In summer each burrow contains only one jerboa, and in winter two jerboas may be found in one burrow.

Some burrows have an emergency exit, which consists of the original hole made in the digging of the burrow, later filled with soil to enable the jerboa to escape through it in time of danger. The emergency exit is shown in types 3 and 4 of *Figure 4*. The presence of a heap of granulated soil a short distance from the opening of the burrow is an indication that an emergency exit is provided.

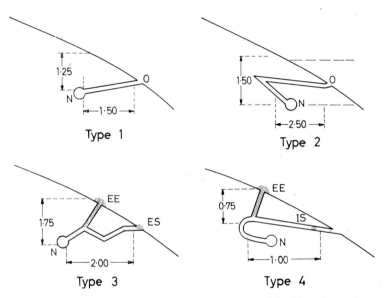

Figure 4. Cross-sections of jerboa burrows: Type 1, straight; Type 2, angular; Type 3, zigzag; Type 4, hooked.
O—opening, EE—emergency exit, ES—external seal, IS—internal seal, N—nest

During the summer months the opening of the jerboa's burrow is usually sealed with soil at the entrance or about 50 cm inside the hole. The bedouin believe that this is to prevent snakes or warm air from penetrating (see types 3 and 4 of *Figure 4*). This soil seal is considered to be a sure indication that the jerboa is in its burrow. In winter the entrance to the burrow is usually left open.

The jerboa's burrow is very clean. No food and no excrements are found in it. The only winter provision that it makes is the jute wool for nesting, which is also found to be clean. All observations indicate that jerboas neither hibernate nor estivate. They are seen, more numerous in winter than in summer, jumping about in the fields an hour or two after sunset until the early hours of dawn. Their muzzling sounds can be heard everywhere.

Jerboa's ecoclimate

The ecoclimate that the jerboa creates in its burrow is completely different from the climate of the desert one or two metres above the ground. In fact, the jerboa lives in two different climates.

Climate above the ground: The climate above the ground represents the natural atmospheric conditions prevailing one or two metres above the ground in desert areas where the jerboa digs its burrow and where it leads a nocturnal life, being active during certain hours of the night. If these conditions are unfavourable, the jerboa does not have to tolerate them except for the short period of time when it is searching for food.

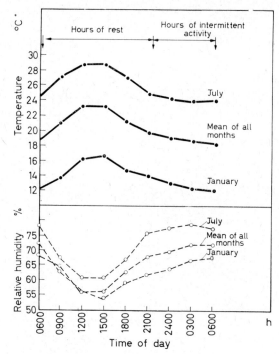

Figure 5. External temperature and relative humidity during the jerboa's daily 24 hours (for Alexandria, 1891–1900).

The external rigorous climatic conditions, in terms of temperature and humidity, which jerboas may have to face during the year are shown in *Figure 5* and in *Table 4* for the months of January (minimum) and July (maximum) (period between 1891 and 1900, according to the Meteorological Report, 1902).

It is perfectly evident from these curves and the table that the jerboa has become nocturnal by necessity. The very hours during which it is intermittently active outside of its burrow, 2100–0600 h, correspond to the hours of the greatest climatic calm. Likewise, the period when the jerboa sleeps in its burrow, 0600–2100 h, coincides with the hours of the day when external climatic conditions are unfavourable and subject to the greatest fluctuations.

Table 4
External climatic conditions

	Time of day (h)							
	0600	0900	1200	1500	1800	2100	2400	0300
Temperature (°C)								
January	12·1	13·6	16·2	16·7	14·8	14·0	13·1	12·5
July	24·2	27·1	28·7	28·8	27·1	25·0	24·4	24·2
Annual mean	18·4	21·0	23·2	23·2	21·2	19·9	19·2	18·7
Humidity (%)								
January	68	64	56	54	59	62	64	67
July	78	67	61	61	67	76	78	79
Annual mean	72	63	56	56	63	68	70	72

Nocturnal life: In fact, nocturnal habits of living present many advantages to a desert animal like the jerboa, whose eyes are adapted to darkness. These advantages are:

(a) decrease in air temperature;
(b) increase in humidity;
(c) reduced air movement;
(d) decreased rate of evaporation;
(e) minimal radiation;
(f) relative ease of finding food;
(g) ease of concealment and escape from natural enemies and danger;
(h) suppression of inhibitions and facilitation of various sensations by the nocturnal calm and light and the existing exceptional conditions.

The microclimate in the burrow: The jerboa can control the temperature and humidity of its burrow in three different ways: (a) by modifying the length and depth of the burrow; (b) by controlling the extent of the soil seal which closes its entrance; and (c) by providing cotton wool for winter months.

During the very hours when the blazing summer sun, with its intensity of radiation, heats the ground and the air above it to intolerable degrees of temperature and dryness, the jerboa's cool and humid burrow provides an ideal place for rest and sleep in the desert. Also, at times when desert storms of wind, sand and rain become extremely trying to living creatures above the ground, the jerboa's burrow is a sure refuge.

Characteristics of burrow microclimate: Williams (1923–1924) has studied the temperature of a jerboa's burrow at Wadi Digla, the desert south-east of Cairo, in August 1922. He recorded the changes of temperature in the jerboa's burrow at a depth of 75 cm and in the shade outside. The temperatures, derived from the curves he published, are as follows:

Outside:	minimum	22·5° C
	maximum	35·0° C
	range	12·5° C
In the burrow:	minimum	29·5° C
	maximum	33·0° C
	range	3·5° C

Analogous observations have been made on other species. Vorhies (1945) reported that even during the hottest season of the year 30° C was recorded in the nest chamber of a kangaroo rat.

Schmidt-Nielsen (1950), who has recorded the temperature and humidity in burrows of desert rodents of Arizona, found that the moisture content of the air in burrows is two to five times higher than that of external air. He concluded that the higher humidity of the burrow reduces the animal's evaporative water loss.

Personal experiments: In order to study the simultaneous temperature and humidity changes in the jerboa's burrow and under outside conditions, the following investigation was made in the Egyptian Western Desert district of Maryut on September 17, 1955.

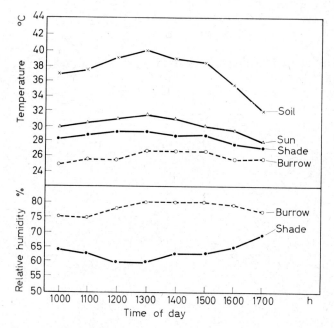

Figure 6. The jerboa's ecoclimate (Maryut, September 17, 1955).

(a) The temperature and relative humidity of the external air, in the shade, were measured with a whirling hygrometer (dry and wet bulb).

(b) The temperature in the sun was measured with the bulb of a thermometer exposed to the direct radiations of the sun 80 cm above the ground, while a light desert breeze was blowing.

(c) The ground temperature was measured with the bulb of the thermometer inserted into the ground.

(d) The temperature and relative humidity of the air in the jerboa's burrow were measured at a depth of 1 m from the surface with a dial thermohygrometer which could be observed through an observation hole for hourly readings.

(e) Water, in milligrams per litre of dry air, was estimated from Bedford's (1948) psychrometric chart.

The results of hourly observations between 1000 h and 1700 h are given in Appendix 4 and illustrated in *Figure 6*. The important points revealed by this study are:

(i) Burrow temperature at its highest level is 2·5° C lower than the air temperature outside, 4·5° C lower than sun temperature and 13° C lower than ground temperature.

(ii) Temperature variations outside affect little the temperature changes in the burrow.

(iii) The relative humidity of the burrow air is 10–20 per cent higher than that of the external air and its variations are more constant in terms of milligrams of water per litre of dry air; the burrow air contains 1·6–5·9 mg more than the external air.

It is evident that the ecoclimate in which the jerboa lives contributes to its body economy in two ways:

(i) The temperature of the burrow is nearer to the comfort zone, and as a result its needs for heat regulation are diminished.

(ii) The higher moisture content in the burrow air reduces its evaporative water loss.

By living in burrows during the day and limiting their activities to the night, jerboas reduce their daily water requirements for heat regulation.

THE JERBOA'S LIFE

Locomotion

The jerboa is a bipedal animal. It uses only its hind-limbs for walking and jumping, whereas its forelimbs serve as a pair of hands for feeding, burrowing, gnawing, grooming, etc.

According to Sanderson artists of Ancient Egypt decorated the murals of temples with images of the jerboa, and hence arose the hieroglyph meaning 'swiftness'.

In fact, by alternate displacement of one foot after the other, or by successive short and long jumps on both feet, or by a combination of the two methods, and by using its tail for equilibrium and counterpoise, the jerboa achieves considerable speeds in locomotion. It can walk forward and backward. Its jumps reach about 2 m in length and 0·5 m in height. Quick to start and quick to stop, its explosive muscular power and its marvellous nervous coordination enable the jerboa to make bird-like springs or 'flight'. It is anatomically well equipped to outrun the enemy and travel long distances in search of food.

Food

The jerboa does not store food in its burrow. It is satisfied with sprouting vegetation when the rainy season is good, or the succulent roots below ground-level, such as scorzonera, when rainfall fails, or the dry grains of barley it digs out of fields when crops fail and there are no plants. It can live without fresh vegetation for a season or more, as dry grains suffice for its subsistence.

Social habits

The bedouin know certain gathering places for jerboas, known as their parlour (Meglis el garabia). This takes the form of a burrow with a larger opening.

At night jerboas come to these 'burrow-clubs' from different directions for playful activity. The dry feces which accumulate at the entrances to these jerboa 'clubs' indicate intense group activity.

Reproduction

Bedouin say that jerboas have two breeding seasons, one in March and the other in July. However, it is possible that there is only one long reproductive period from March to July. Jerboas give birth to 4–7 young. Actually, two pregnant females brought to the laboratory in March 1958 gave birth to four young each. But in two burrows which we opened in July 1954, seven young and the mother and six young and the mother were found.

The female jerboa's burrow is longer and deeper than that of the male. She is usually found in the burrow with her young during the breeding and suckling season.

Enemies

The jerboa's burrow is also its refuge from enemies, which include the snake, desert fox, fennec, etc., and also Man.

Bedouin hunt the jerboa, for they like its roasted meat. They catch it by pouring water into its burrow, which obliges the jerboa to rush out, or by gradually digging down the burrow towards its abode and capturing it there, or by setting overnight traps near the burrows in which several jerboas are caught during their nocturnal wanderings. The traps consist of improvised stone walls with a net in the middle and a pit underneath this.

The bedouin are acquainted with the habits and manners of jerboas, and many fables, such as the one narrated by Musil (1928), indicate that they consider it to be an intelligent and wise animal.

CONCLUSIONS

1. The jerboa (*Dipus aegyptius* or *Jaculus orientalis*) studied here is a desert rodent of the family of *Dipodidae*, North African.

2. A type of rodent, of Asiatic origin, similar to *D. aegyptius* must have invaded North Africa in the distant past, and, with the advent of desert conditions, it must gradually have evolved into its present form.

3. *D. aegyptius* or *J. orientalis* is a larger and more robust animal than the *J. jaculus*. *D. aegyptius* is adapted to desert conditions along the littoral zone of North Africa, west of Alexandria, whereas *J. jaculus*, the smaller jerboa, inhabits desert areas south of the littoral zone where the climate is hot and dry.

4. The jerboas studied here were captured in their burrows in the Egyptian Western Desert littoral zone near Alexandria, where the climate is temperately warm, and a two-seasonal periodicity of a short rainy winter

and a long rainless summer prevails. However, periods of drought lasting several years are known, when rainfall is insufficient for vegetation.

5. The jerboa is anatomically well equipped for desert life. All parts of its body (head, trunk, hands, feet, tail, colouring of the fur) are adapted to the type of life it leads. It is a burrowing, nocturnal, explosively muscular and swift-moving rodent.

6. The jerboa creates, in the burrow it digs, an ecoclimate of its own in the midst of the desert. This ecoclimate serves the jerboa in three ways: (a) to live under climatic conditions nearer the comfort zone, and thus escape the adverse climatic conditions outside; (b) to economize water and energy, as its daily metabolic requirements are reduced; and (c) to rest and sleep in a shelter which gives protection against its natural enemies; namely, stormy weather, animals and Man.

In short, underground living and nocturnality are the partial solutions which jerboas have found to the three main problems of the desert—scarcity of water, rarity of food and intensity of heat and solar radiation.

4
THE JERBOA IN THE LABORATORY

At the beginning of this study, we attempted to imitate the jerboa's burrow by constructing a box cage 90 cm long, 15 cm deep and 13 cm wide. An opening of 8 cm² was provided at each end of the box. The top had two covers, the lower of glass and the upper of wood. By opening the upper wooden cover the jerboas could be observed without disturbing them. This box cage was placed in a large tray filled with sand. At the corners of the tray we placed food receptacles. Pieces of wood and cloth were also provided.

Jerboas lived comfortably in this cage, or pseudo-burrow, for some time, and their natural activities could easily be observed. From the way they accepted their new habitat, it was evident that five things were essential for their well-being, namely: (a) food; (b) a cage for hiding, resting and sleeping; (c) pieces of wood for gnawing and wool for nesting; (d) sand to imitate the digging of the burrow and to play with; and (e) company of other jerboas. This imitation burrow seemed to satisfy all their needs.

NUTRITION

Barley, wheat and lettuce are the preferred foods of jerboas. Moreover, they eat bread, rice and vegetables, such as fresh corn, green peas, green beans, carrots, potatoes, lentils, etc. (see *Figure 7*). They also eat peanuts and melon seeds. They do not eat dates, dry fruits, bananas or tomatoes.

Jerboas eat with their forelimbs, which serve as little hands. Holding the barley grain between their two hands, they manage to peel it completely with their teeth in order to eat the interior. With wheat, they break the grain into halves and gnaw the interior.

Jerboas habitually eat in the evening and dawn. They do not soil their imitation cage. They defecate and urinate around the areas where they spend the evening, eating and playing.

Jerboas drink water when they cannot find green plants. However, there were cases in the laboratory when the jerboas voluntarily abstained, for some time, from eating lettuce and drinking water in preference to dry grains of barley and wheat. The important fact is that, in general, all jerboas have the capacity to live only on dry grains of barley and wheat for periods of more than one year.

THE PSEUDO-BURROW

In the pseudo-burrow described above jerboas create the ecoclimate (microclimate) of their preference by controlling the size of the openings. They manage to block one or the other of its openings (half, three-quarters or entirely) by means of a mixture of sand, pieces of wood and wool in order to change the ventilation of their habitat. During the hours when they are

THE PSEUDO-BURROW

naturally active, they can be seen preparing the shelter of their choice with highly skilled and coordinated movements of all the parts of their body.

The following experiment indicates clearly that jerboas prefer a humid atmosphere similar to that to which they are accustomed in their natural

Figure 7. A jerboa eating.

Table 5

The jerboa's preferred ecoclimate

Date	Time (h)	In the laboratory		In the cage	
		T (°C)	R.H. (%)	T (°C)	R.H. (%)
29.9.52	11·00	25·6	67	26·1	72
	20·00	25·0	68	25·6	74
30.9.52	10·30	25·3	69	25·5	74
	22·30	25·0	70	25·0	75
2.10.52	10·30	25·4	74	25·4	81
	17·00	25·6	77	26·6	82
3.10.52	10·30	25·0	78	25·0	86
	13·15	25·5	79	25·5	83
4.10.52	10·30	25·0	75	25·0	84

burrows. By means of two thermohygrometers, one in the laboratory and the other inside the pseudo-burrow, where six jerboas lived together, the data of *Table 5* were recorded.

It is to be noted that the temperature is often the same in the room and in the cage; the relative humidity varies similarly but is higher in the cage where the jerboas live, as would be expected.

This observation on the climatic preference of jerboas in an artificial habitat confirms the opinion of Russell (1949), who believed that the instinct of self-preservation drives animals to discover and maintain their normal ecological conditions.

REST AND SLEEP

All day long, from dawn till dusk, jerboas remain in their communal cage, sleeping most of the time. They like to sleep on top of each other. Their sleep is profound and their relaxation is complete. They often change position during their sleep. When they are not sleeping soundly they assume, from time to time, a crouching position (see *Figure 8*); the body rests on the hind-limbs and tail, while the head and back are bent forward towards the abdomen, the nose touching the lower part of the abdominal wall. In this spherical form, only the animal's fur, ears and tail can be seen.

Figure 8. A jerboa in the crouching position.

This crouching position of semi-rest enables the jerboa, during the winter, to warm and humidify the cold environmental air that it breathes, bringing it close to its body temperature. It also serves to warm the animal by the hot breath coming out of the respiratory tract, while reducing the surface area of its body exposed to cold. During the summer, the spherical form enables the jerboa to heat and humidify, close to the temperature of its skin, the environmental dry air which it breathes. Thus, the crouching helps to reduce heat and water losses during the winter, and water loss by evaporation (insensible perspiration) during the summer: it is a mechanism of heat and water conservation in winter, and water conservation in summer. The jerboa therefore creates a microclimate close to its abdomen by means of the air it expires—a sort of air-conditioning—in order to be able to fight more effectively against cold and dryness (see *Figure 8*).

BEHAVIOUR

In its cage, or pseudo-burrow, the jerboa carries out its habits of constructing a shelter as in nature. One can observe in the laboratory its dexterity in preparing a burrow, its astonishing mastery and rapidity in making wool from a piece of cloth, its habit of gnawing everything that can be gnawed and its skilful use of hands and hind-limbs in manipulating sand and every other material available in order to prepare a small corner as its refuge.

Graceful, playful and peaceful while in security, it is capable of defending itself ferociously when it is menaced. There were certain cases of fierce quarrels among jerboas in the laboratory. If a timely intervention is not made the chase between two quarrelling jerboas continues, savage cries fill the air and sometimes other jerboas join the combat, which ends with one or several jerboas being severely mutilated by bites. But these cases of savage conduct are rare and, in general, jerboas live together peacefully. They make muzzling sounds to show their annoyance and anger.

The jerboa is a sociable animal: it seeks the company of other jerboas. Given spacious surroundings, it will spend hours playing, running and jumping. In spite of its timidity and fearfulness, it quickly becomes accustomed to human attention. It is curious and likes to play and be caressed. It keeps itself clean and has regular habits of grooming. Its excrements have no offensive odour.

The jerboa will tolerate the presence of a white rat in its cage, but remains indifferent towards the intruder.

From time to time jerboas practise rhythmic tapping and scratching movements with their hind-limbs against the floor of their cage. This tapping may last more than a quarter of an hour and gives the impression of being a form of communication. As these movements resemble those of burrow-digging, it may be that the jerboa imitates the digging movements which it misses while living in cage. This rhythmic activity may, however, be an habitual pastime of the jerboa. According to Bourlière (1951), other rodents (*Meriones shawi, Pachyuromys duprasi*) also practise tapping on the soil.

IDENTIFICATION

After an initial period of general observation on the life and behaviour of jerboas in the laboratory, it became necessary to find a way of identifying the animals in order to begin controlled experiments.

The first requirement was to determine the type of cage that would serve the jerboa's needs. After several trials, the cage described in Appendix 1 was devised. The jerboas were introduced into the laboratory in separate cages, either singly or in pairs. Each cage was numbered. Also, a large cage was constructed to provide freedom of movement to the occupants of small cages, each in its turn.

In order to identify jerboas individually, the following method was adopted: Each jerboa brought to the laboratory received a serial number, and its weight, sex and the date and locality from which it was captured were recorded. Also, each jerboa was marked on the tuft of its tail. The tuft is 8 cm long, in two colours, comprising 3 cm of black hair followed by 5 cm of white hair laterally spread on both sides of the tail axis and at its extremity. By clipping the black or white parts of this flat tuft on either side, it was possible to make eight marks easily recognizable at first sight. These marks can be made as follows: the jerboa is placed in the weighing-box, with it tail protruding, and parts of its hair are clipped off quickly and without disturbing the animal. This operation is repeated every two months. These marks, combined differently, permit the identification of more than thirty jerboas.

Thus, each jerboa is recognized by its identification mark, in addition to its serial number. For example, J2RB indicates jerboa No. 2, which has the black hair cut off the right side of its tuft.

For a small research laboratory this system of identification is more practical than the classical method of notching the ears of the animal.

A jerboa can be handled by the tail without causing disturbance, if the transfer from one cage to another is made rapidly. If the transfer from one cage to another is not pressing, it is better first to let the animal enter the weighing-box spontaneously and then make the transfer without touching it.

In order to identify jerboas' sex, the following technique was devised: Holding the root of the jerboa's tail with the left hand, its abdomen is turned towards the right hand. In this position, the animal remains quiet, with its hind-limbs drawn towards its abdomen and its sexual organ exposed. Separating the hairs of the anal region with the index-finger of the right hand, the male or female sexual organ of the animal can easily be identified. As mentioned previously, the male *Dipus aegyptius* has at the dorsal part of its glans penis two horny spines (characteristic armature).

REPRODUCTION

Jerboas do not reproduce in the laboratory. Several couples were kept together, for long periods, without results. The problem of reproduction in captivity necessitates a special investigation. According to Hediger (1953), many species of animals do not reproduce in captivity, or do so with difficulty.

Nevertheless, we have observed that female jerboas impregnated in the desert had a normal gestation during their captivity. It is, therefore, the fecundation by the male that does not take place.

CONCLUSIONS

The jerboa is a nocturnal rodent, sociable and amenable to life in captivity if certain types of cage and material are put at its disposal. It is a clean animal, easy to feed and economical to keep. Unfortunately, the essential conditions for its reproduction in the laboratory could not be realized. If this problem were successfully solved, a complete study of this species could be undertaken.

5
COMPARATIVE GROWTH OF THE JERBOA AND RAT: DIET AND BODY WEIGHT

GROWTH

Growth of jerboas and white rats

A pregnant jerboa captured in its habitat and brought to the laboratory gave birth to four young on March 20, 1958. We have made regular observations on the growth of two of these jerboas. At the same time we observed two rats, of the Wistar breed, born in the laboratory on April 1, 1958. These growing jerboas and rats lived under the same climatic conditions and received the same classical diet. The results of our observations appear in Appendix 5.

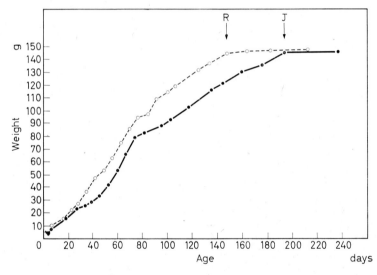

Figure 9. Growth curves of jerboas (———) and rats (-----). Arrows indicate attainment of maximum body weight.

The comparative growth curves (*Figure 9*) indicate that on the basis of weight increase in relation to age, jerboas grow more slowly than rats. The rat reaches its maximum body weight in 147 days, whereas the jerboa reaches it in 193 days.

Body development

Our observations on the development of the above animals gave the following results.

The period of the jerboa's infancy—namely, its dependence on maternal help—is twice as long as that of the rat. The jerboa, like the rat, is born naked, without hair, with its ears closed. But it takes double the time for the jerboa's hair to appear and the ears to open compared with the rat. Also, born blind, like the rat, the jerboa's eyes do not open until the 36th day, whereas the rat's eyes open by the 12th day.

The jerboa, in comparison with the rat, is slow in the development of its forelimbs and hind-limbs. Whereas the rat runs and controls its movements by the 15th day, the jerboa does not acquire the complete usage of its

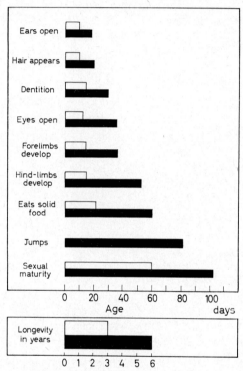

Figure 10. The development of the jerboa and white rat: ■ jerboas; ☐ white rats.

hands and hind-limbs until the 53rd day. Even then it is not able to jump until 81 days old.

The sexual maturity of rats is known, as the time when they start to reproduce can be determined. According to Griffith and Faris (1942), rats, male and female, reach sexual maturity between the 50th and 60th days of their age. As jerboas do not reproduce in captivity, it cannot be established when the female jerboa reaches sexual maturity. But for the male jerboa the outgrowth of the two horny spines at the dorsum of the head of the penis serves as an indication of sexual maturity. On the basis of this criterion, the male jerboa can reproduce by the 102nd day of its age.

GROWTH

The jerboa lives twice as long as the rat. We have observed that even in captivity the jerboa's life duration exceeds six years (three years for the rat).

Figure 10 shows the results of the comparative study, mentioned above, on the development of jerboas and white rats.

Body weight and sex

Two groups of young jerboas with their mothers, brought to the laboratory in 1954 from two different burrows, enabled us to study the influence of sex on body-weight variations during growth and after maturity.

Six jerboas of the same litter, three males and three females, were put under observation. Appendix 6 gives the results of this study. From the averages of this appendix *Figure 11* was prepared, showing curves for males and for females. These indicate that the body weight of the male jerboa is slightly higher and increases more rapidly than that of the female jerboa. At maturity, the weight of males is in general higher than that of females.

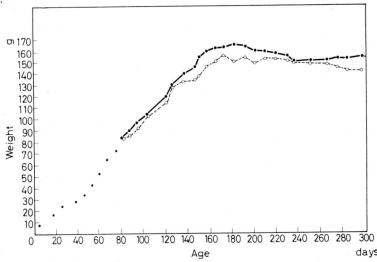

Figure 11. Body-weight variations of male (———) and female (-----) jerboas (mean values of three males and three females of the same litter; global weights until 80th day).

It is fitting to remark here that rats of the same age but different sex differ greatly in weight; the adult male rat reaches 300 g while the female rat reaches only 150–170 g.

Conclusion

The jerboa's growth curve is similar to that of the Wistar rat, but, unlike rats, male and female jerboas have growth curves that are very much alike.

NUTRITION AND BODY WEIGHT

The effect of wet diet (barley, wheat and lettuce) and dry diet (dry grains of barley and wheat) administered *ad libitum* on the weight variations of jerboas was studied as follows.

Effect of dry diet on body weight of adult jerboas

Two jerboas, a male and a female, were put on dry diet when they reached 191 days of age. Appendix 7 and *Figure 12* give the results of this investigation. The curves indicate that the change of nutrition from wet diet to dry diet had no effect on the maximum reached. In fact, instead of being harmful, the dry diet has caused a slight increase in weight.

Effect of chronic dry diet on body weight of jerboas

The results of this research appear in Appendix 7. Our findings indicate that jerboas have the capacity to live on dry diet for periods varying from one year to three years or more.

Jerboa J51 managed to live for more than three years on dry grains of barley and wheat only (containing 11–12 per cent water).

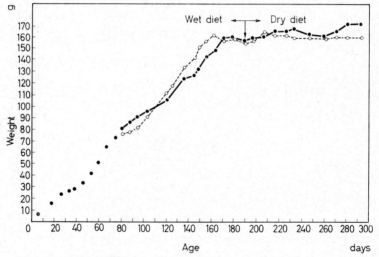

Figure 12. Effect of dry diet on body weight of adult jerboas: ——— J46, male; – – – – J41, female.

There is no doubt that the mechanism which suffers when dry diet is prolonged is that of renal elimination. The animal reaches a point where it refuses to eat, and its body weight diminishes rapidly. This can be remedied simply by administering wet diet before its body weight drops to 100 g, or a loss of about 40 per cent.

Our findings on Jerboa J38, which appear in Appendix 8, prove this important point. Change of nutrition from dry to wet diet at the critical moment when the animal's body weight dropped to 99 g helped not only to recover lost weight but also to exceed it by reaching 201 g.

This experiment indicates that dry diet causes no irreversible organic lesion, if the intervention comes in time. What is deranged is the functioning of the urinary mechanism owing to the accumulation in the blood of the waste products of metabolism that the animal is unable to excrete. The maximum concentration of urea elimination cannot surpass 120–130 g per 1000 cm^3 (Le Breton; Jacquot; Spector, 1956): above this the animal will probably die of uremia. This is why it refuses to take dry food. Once the water that is necessary for purging the kidneys is provided through wet diet, the functioning of the urinary mechanism becomes normal; the animal begins to eat, and it is saved from its critical state within a week.

The results of our experiment on Jerboa J63, as given in Appendix 8, furnish additional evidence in support of what is said above. After nine months on wet diet, it was put on dry diet. At the completion of ten months on dry diet, its body-weight started to drop. Restored to wet diet, its weight reached a record of 196 g. This experiment proves that the non-exaggerated dry diet, which obliges the kidneys to do sustained and abnormal work of concentration, is not 'toxic' for an animal whose parenchyma permits this work for a long time (Ambard, 1914). The well-balanced dry diet is, therefore, a normal diet for jerboas but it is lethal to rats, as will be explained in the following chapters.

Post-mortem examination of lethal cases among jerboas on prolonged dry diet has shown a diminution of body weight to 90–100 g (because of inanition), anemic and granulated liver, as in cirrhosis, ballooned intestines and, particularly, the presence of venous congestion in the cortical renal region.

It is evident that when the functioning of the kidneys is deranged, the animal ceases to eat and its body weight diminishes. Death results from deficient urea elimination and from hyperthermia provoked by a defective thermoregulation.

Conversely, when the rat is put on dry diet it begins to eat less each day for the first three days, and then completely ceases to eat. As a result, its body weight diminishes very rapidly, and after a few days it dies.

In short, the rat does not tolerate dry diet for more than three days, whereas the jerboa can tolerate it for three years. The difference between the two animals on dry diet is attributable to their different renal functions.

CONCLUSIONS

1. The period of the jerboa's bodily development is twice as long as that of the rat, and its longevity is double that of the rat.

2. The jerboa possesses physiological mechanisms which permit its adaptation to desert conditions, especially as its water needs are limited.

3. The structure and function of the jerboa's urinary system play an important role in its capacity to live for years on only dry grains of barley and wheat.

4. The jerboa tolerates a diet very poor in water (10 per cent) for long periods (1–3 years). Death is preceded by a short and sudden anorexia. The rat cannot survive such a diet, except for a short time.

5. Experimental data, presented in the following chapters, permit the conclusion that the renal functioning (maximum concentration of urea) and thermoregulation mechanisms characteristic of the species jerboa explain the facts noted.

6

EFFECT OF DIET ON EXCRETION AND BODY WATER CONTENT

INTRODUCTION

The relation of nutrition to excretion is well known. The animal receives its food and water from the environment and returns the waste products of its metabolism to the environment. Scarcity of food implies the diminution of excrements.

The problems of nutrition and excretion have been extensively investigated by physiologists. Adolph (1933) studied the metabolism of water and its distribution in the tissues of the body. Schmidt-Nielsen and his associates (1948) found that the *Heteromyidae* (the kangaroo rats of the Arizona desert) are able to live on dry grains indefinitely. They can maintain the water content of their body at 66·4 per cent (fats included), and concentrate their urine more than other mammals. B. and K. Schmidt-Nielsen (1951) also fed kangaroo rats and white rats on pearled barley. As a result, the water content of the feces of kangaroo rats was 45 per cent, whereas that of white rats was 68 per cent. This means that for 100 g of pearled barley consumed the water lost through feces by the kangaroo rats is 2·54 g, whereas that lost by the white rats is 13·5 g (five times as much).

Another factor closely related to food consumption and excretion is the natural activity of the animal. It is known that the more active an animal is the more it needs energy and water. Therefore, to explain why certain animals consume more food than others, it is necessary to investigate their normal daily activity level.

The 24-hour activity level of a great number of small animals has been investigated. Browman (1952) reported that the 24-hour cycle of voluntary activity of white rats could be altered to a 16-hour one by subjecting the animals to an experimental training in an environment where light and temperature are controlled.

NUTRITION AND EXCRETION OF THE JERBOA

From the first days that jerboas were brought to the laboratory, it was observed that they consumed less food and excreted less feces and urine than white rats of the same weight. In order to study more precisely the effect of nutrition on the excretion of feces and urine in jerboas, we have conducted the following experiments.

Experiments

Twelve jerboas (*Group I*) were kept on a wet diet composed of wheat, barley, lettuce and water for a period of more than one year.

Two jerboas (*Group II*) were kept on a dry diet composed of wheat and barley for a period of more than one year.

Method: The period of the year chosen for the experiments was August–September, when the environmental temperature and humidity vary very little.

The metabolism cage described in Appendix 1 was used.

Jerboas of Group I were fed on wet diet composed of wheat and lettuce for a period of six days, after which they were fed on dry diet composed of wheat grains only for another period of six days. The body weight, food consumption and excretion of feces and urine were measured each day at the same hour.

Jerboas of Group II were fed on dry diet, as usual, composed of wheat grains only, for two periods of six days each. The body weights, food consumption and excretion of feces and urine were measured and recorded each day at the same hour.

In this series of experiments, the feces and urine of the animals were measured in the state in which they were collected every 24 hours, without taking into account the common factor of evaporation.

Results and interpretation

The results of this study are presented in Appendix 8; *Figure 13* shows them in graphic form. The findings indicate the following:

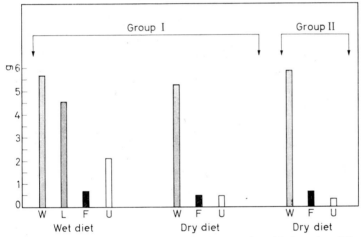

Figure 13. Effect of wet and dry diets on the excretion of jerboas (per 100 g body weight per day).

W—wheat, L—lettuce, F—feces, U—urine

Temporary effect of dry diet: When jerboas of Group I, accustomed to wet diet, are put on wet diet for six days, they consume 5·73 g of wheat and 4·58 g of lettuce, and excrete 0·69 g of feces and 2·10 cm³ of urine per 100 g of body weight per day. Of the 12 jerboas of Group I, only one jerboa, J27, voluntarily abstained from eating lettuce, preferring a dry diet. As this is an exceptional case, its measurements were not included in the average of Group I.

When the same jerboas of Group I, accustomed for one year to wet diet, are put on dry diet for six days, they consume 5·31 g of wheat and excrete 0·5 g of feces and 0·47 cm^3 of urine per 100 g of body weight per day. This means that they eat less and excrete less. Their urine becomes concentrated and acidic in reaction; namely, from pH 8 (on the basis of the universal pH indicator by Curr) on wet diet, it falls to pH 5 on dry diet. Moreover, it is very significant to note that the average of their body weight is practically unchanged (158 g on wet diet and 157·1 g on dry diet) in spite of the fact that they eat less during the six days of dry diet. This means that the jerboas can live on wet diet as well as on dry diet by adjusting their metabolic processes so that water ingested equals water lost.

Chronic effect of dry diet: When jerboas of Group II, which lived for more than one year on dry diet, are put on the same dry diet for 12 days, they consume 5·92 g of wheat and excrete 0·64 g of feces and 0·31 cm^3 of urine per 100 g of body weight per day. Their food consumption and excretion do not differ very much from those of jerboas of Group I temporarily put on dry diet. The significant chronic effect of dry diet is a greater concentration of urine, accompanied by a more acidic reaction of urine (pH 3). The jerboa can, therefore, live on dry diet for a long time, by reason of its capacity to continue eating sufficiently and excreting urine which is more and more concentrated in urea and salts and which is acid in pH.

EFFECT OF DIET ON EXCRETION OF JERBOAS AND RATS

The findings of the preceding experiments led us to study in greater detail the effect of diet on the excretion of jerboas and white rats, at the environmental temperature of 25° C.

Experiments

In order to be able to compare, under identical experimental conditions, the influence of wet and dry diets on the excretion of jerboas and rats, the experiments described in Appendix 1 were conducted.

The results of these experiments appear in Appendix 8. *Figure 14* shows these results per 100 g of body weight per day at the environmental temperature of 25° C.

Results and interpretation

Wet diet: The results of these experiments confirm those of the experiments of the preceding paragraph. They indicate that, at the environmental temperature of 25° C and per 100 g of body weight per day, jerboas consume 5·960 g of wheat and 4·892 g of lettuce, whereas rats consume 6·385 g of wheat and 16·365 g of lettuce. Similarly, the excrements of jerboas are less than those of rats. Jerboas excrete, per 100 g of body weight per day, 1·009 g of feces (fresh) and 2·949 g of urine (fresh), whereas rats excrete 2·562 g of feces and 5·127 g of urine.

Dry diet: The striking difference between jerboas and white rats is the capacity of jerboas to live on dry diet by reducing their consumption of wheat to 3·887 g per 100 g of body weight per day at the environmental temperature of 25° C, whereas rats cease to eat after the third day on dry diet. This shows that dry diet becomes lethal to rats because of inanition as well as water fasting.

Also, from the point of view of excretion, jerboas have the capacity to reduce their excrements to 0·608 g of feces and 0·692 g of urine per 100 g of body weight per day. But rats excrete very little feces and urine the first day, and hemorrhagic symptoms appear, because of the dry diet, towards the sixth day, indicating that irreparable lesions are inflicted on the urinary and intestinal tracts.

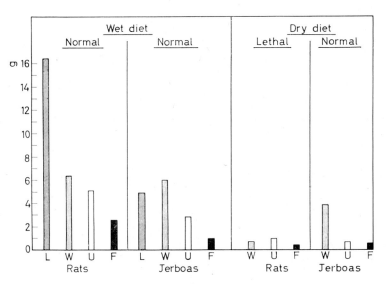

Figure 14. Effect of wet and dry diets on the excretion of jerboas and white rats (per 100 g body weight per day).
L—Lettuce, W—wheat, U—urine, F—feces

In conclusion, dry diet is normal for the jerboa, since it has the capacity of subsisting on it for several years without undergoing any organic impairment. The rat, however, cannot bear dry diet for more than three days. It ceases to eat because it is unable to eliminate the waste products of its metabolism by means of its excretory mechanisms, which fail.

WATER CONTENT OF THE BODY

Brody (1945) found that the body of adult animals, without their fatty tissues, contains about 70 per cent water. As the fat depots contain 6–20 per cent water, whereas the other tissues contain more than 60 per cent, by deducting the weight of fatty tissues from the total body weight it is possible to establish more precisely the water content of the body. Babinau and Pagé (1955) reported that the non-fat dry matter of rats is proportional to the total water content of the body.

Experiments

The post-mortem examination of several animals in the laboratory

EFFECT OF DIET ON EXCRETION AND BODY WATER CONTENT

Table 6

Percentages of water and fat in the bodies of jerboas and rats

	Initial body weight (g)	Body weight dry (g)	Total body fat (g)	Fat (%)	Fat-free dry weight (g)	Fat-free dry weight (%)	Body weight without fat (g)	Body water (g)	Water content (%)
White rats, wet diet (11.4.58)									
R28F	159·0	54·50	18·53	11·6	35·97	22·6	140·46	104·49	74·3
R27M	224·5	78·43	26·66	11·8	51·76	23·0	197·83	146·07	73·8
R23M	242·5	79·62	20·70	8·5	58·92	24·3	221·79	162·87	73·4
Average				10·8		23·3			73·8
Jerboas, wet diet (7.1.59)									
J72F	168·5	77·54	52·72	31·3	24·81	14·7	115·77	90·96	78·5
J70M	184·3	87·25	59·33	32·2	27·92	15·1	124·97	97·05	77·5
Average				31·7		14·9			78·0
Jerboa, dry diet (7.1.59)									
J51F	144·32	56·83	26·14	18·1	30·68	21·2	118·17	87·49	74·0

revealed that the body of the jerboa on wet diet is fat, whereas that of a jerboa on dry diet is lean. We have often found masses of fatty tissue surrounding the organs in the abdominal cavity of jerboas on wet diet. But the bodies of jerboas on dry diet were lacking this supplement of fats.

As it is important in this study to know the percentage of water and fat in the body, the method described in Appendix 1 was adopted. The experiments conducted on three rats and two jerboas on continuous wet diet, and one jerboa on continuous dry diet, gave the results given in *Table 6*.

Results and interpretation

The results indicate that the body water content of jerboas on wet diet (fatty tissues excluded) is 78·0 per cent, whereas that of the rat is 73·8 per cent. The fat content of the jerboa's body is 31·7 per cent, whereas that of the rat is 10·8 per cent. These contents, of 20·9 per cent more fat and 4·2 per cent more water in favour of the jerboa, are important when considering thermoregulation. Even the jerboa on dry diet has 7·3 per cent more fat and 0·2 per cent more water in its body than the rat on wet diet.

Furthermore, the jerboa on dry diet has 13·6 per cent less fat and 4 per cent less water in its body than jerboas on wet diet. The chronic effect of dry diet is a diminution of fat and water in the body. This is why the jerboa on dry diet is less resistant to high environmental temperatures than one on wet diet, as it lacks reserves of water and does not ingest any water.

The fat-free dry body weight, as expressed in percentage of the initial body weight, is also significant, namely: rat on wet diet, 23·3 per cent; jerboa on wet diet, 14·9 per cent; jerboa on dry diet, 21·2 per cent.

SPONTANEOUS ACTIVITY AND NUTRITION

What is the effect of diet on the spontaneous activity of animals? In an effort to find the answer to this question, we have devised an actograph whose description is given in Appendix 1.

Tracings of the natural activity of several rats and jerboas were recorded in the laboratory. *Figure 15* shows an example. These are the tracings, during two consecutive days, of the 24-hour activity cycle of white rat R19, on continuous wet diet, of jerboa J47, on continuous wet diet, and of jerboa J42, on continuous dry diet.

From an examination of these tracings, and similar ones of other animals studied in the laboratory, the following observations are made:

1. Each species of animal has movements characteristic of its individual constitution.

2. Rats on a continuous wet diet remain active during 24 hours, with short periods of reduced activity every 2–3 hours.

3. Jerboas on a continuous wet diet show an excessive activity between 2100 h and 0600 h, a reduced activity between 0600 h and 0800 h, a period of complete rest between 0800 h and 1800 h and a reduced activity between 1800 h and 2100 h.

4. Jerboas on a continuous dry diet show an intermittent activity between 2100 h and 0700 h, a period of interrupted rest between 0700 h and

1600 h, a reduced activity between 1600 h and 1800 h, followed by a period of interrupted rest until 2100 h.

In general, the tracing of the jerboa on dry diet indicates a feeble activity compared with that of jerboas and rats on wet diets.

In short, the tracings of the animals studied clearly indicate that a dry diet obliges the jerboa to reduce its natural movements considerably. Its period of rest becomes longer, but intermittent. The form of the rat's activity is different from that of the jerboa. The rat remains active during 24 hours, without a prolonged period of rest, whereas the jerboa has a period of complete rest which lasts more than nine hours.

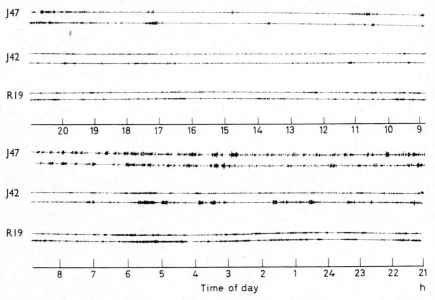

Figure 15. Tracings of the 24-hour natural activity cycle of jerboas and white rats (two consecutive days).

J47—jerboa on wet diet, J42—jerboa on dry diet, R19—white rat on wet diet

CONCLUSIONS

Effects of wet diet

1. The jerboa eats less than the rat. Per 100 g of body weight at the environmental temperature of 25° C, the jerboa consumes 5·760 g of wheat and 4·982 g of lettuce, whereas the rat consumes 6·385 g of wheat and 16·365 g of lettuce.

2. The excrements of the jerboa are less than those of the rat. Per 100 g of body weight at the environmental temperature of 25° C, the jerboa excretes 1·009 g of feces and 2·949 g of urine, whereas the rat excretes 2·562 g of feces and 5·127 g of urine.

3. The body-water content of the jerboa (fatty tissues excluded) is 78·0 per cent, whereas that of the rat is 73·8 per cent. Also, the body-fat content of the jerboa is 31·7 per cent, whereas that of the rat is 10·8 per cent. The

fact that the jerboa has 4·2 per cent more water and 20·9 per cent more fat in its body than the rat plays an important role in thermoregulation, particularly in its resistance to high environmental temperatures, which, *a priori*, seems paradoxical.

4. The spontaneous activity of the jerboa during the daily 24 hours is less than that of the rat. The jerboa has a long period of complete rest during the day, whereas the rat remains active with short periods of reduced activity. This is one of the reasons why the jerboa eats less—it spends less energy in muscular work during the day.

Effects of temporary dry diet

A jerboa put on a temporary dry diet eats less and excretes less. Its urine becomes concentrated and acidic in reaction, which is one of the characteristics of this small rodent. Its weight diminishes a little. It manages easily to adjust its metabolic processes to a dry diet. However, a rat put on temporary dry diet ceases to eat after the third day. Its body weight diminishes rapidly and towards the sixth day lesions appear in the urinary and intestinal tracts. Dry diet becomes lethal to the rat, whereas the jerboa accepts it easily because of the characteristics of its urinary secretion.

Chronic effects of dry diet

1. The jerboa can live on a dry diet for 1–3 years by eating little and excreting little. At the environmental temperature of 25° C and per 100 g of body weight per day the jerboa on continuous dry diet consumes 3·887 g of wheat and excretes 0·608 g of feces and 0·692 g of urine.

2. A jerboa on a continuous dry diet has 13·6 per cent less fat and 4 per cent less water in its body than a jerboa on wet diet. This diminution of fat and water in the body of the jerboa also diminishes its capacity to resist high environmental temperatures.

3. A jerboa on a continuous dry diet has 7·3 per cent more fat and 0·2 per cent more water in its body than a rat on wet diet.

4. The 24-hour natural activity cycle of the jerboa on continuous dry diet is feeble compared with that of jerboas and rats on wet diet. Dry diet obliges the animal to diminish its spontaneous activity in order to reduce its consumption of energy and water.

REGULATION OF BODY TEMPERATURE

BODY-TEMPERATURE VARIATIONS

Body-temperature variations in relation to thermal environment have served in the past as a basis for the classification of animals into three categories:

  (a) Cold-blooded animals, poikilotherms, whose body temperature changes with the temperature of the environment;
  (b) Heterothermic animals, whose temperature regulation is limited;
  (c) Warm-blooded animals, homotherms, which regulate their body temperature and maintain it constant, within certain limits, in relation to external temperature variations.

There is no doubt that animals have developed in the course of evolution certain thermoregulatory mechanisms. According to Allee and collaborators (1950), of the million or more known animal species only 20,000 are homotherms: birds and mammals belong mostly to this category.

Gulick (1937) has reported that the full development of homothermy in rats takes place during the period of suckling (20 days). According to Brody (1945) and Spector (1956), the rat's body temperature is 37·3° C, but it may attain 38·1° C.

Cycles

The body temperature of homotherms is not constant throughout the daily 24 hours but it follows a curve of variations presenting a maximum. The range between minimum and maximum temperatures is more marked in certain species (birds).

Burckhardt, Dontcheff and Kayser (1933) have attributed the 24-hour cycle in the body temperature of the pigeon to visual and auditive excitations. Chevillard (1935) has found that the daily cycle in the body temperature of small rats reaches its maximum between 2100 and 2400 h and its minimum between 0900 and 1200 h, corresponding to the normal feeding activity and rest period of these animals. Welsh (1938), in his review of the diurnal rhythms in biology, has confirmed that the body-temperature variations reflect the cycle of activity.

Schmidt-Nielsen (1958) has found that the rectal temperature of the camel deprived of water fluctuates between 35° C early in the morning and 41° C at 1200 h, i.e. 6° C. But as soon as the camel drinks water the range of fluctuation becomes narrower.

BODY TEMPERATURE OF THE JERBOA AND RAT

In reviewing the scientific literature on body temperature, we have found no mention of the jerboa's temperature. Therefore, the questions which required experimental investigation were:

  (a) What is the body temperature of the jerboa and how does it compare to that of the white rat?

(b) How do jerboas and rats regulate their body temperature under environmental temperatures of 0–40° C?

(c) Does the body temperature of jerboas vary with the ingestion of dry diet?

The preliminary tests which were undertaken revealed the following notable facts:

We have seen that the jerboa's movements have an explosive character and are therefore accompanied by sudden production of heat, which, as would be expected, affects the body temperature. This extraordinary capacity to accelerate heat production instantly is paradoxically coupled with its capacity to relax completely. At rest, the jerboa is in total muscular relaxation. It can suddenly fall into a deep sleep as easily as it can arouse into intense activity.

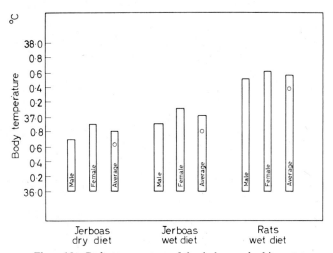

Figure 16. Body temperature of the jerboa and white rat.

These general observations on the jerboa led us to conceive a uniform method of taking its rectal temperature, so as to avoid fluctuations owing to this great excitability. This method is described in Appendix 1.

Experiments

Rat's temperature: In order to compare the jerboa with the rat, it was deemed necessary to make a series of body-temperature readings of Wistar rats, which lived in the laboratory under environmental conditions identical to those of the jerboas.

Appendix 19 shows that these white rats had an average body temperature of 37·5° C. This result verified what had previously been observed by other authors.

Jerboa's temperature: By use of the same method, the body temperature was investigated of eight jerboas on wet diet and eight on dry diet. Only when the animal was absolutely quiet were the temperatures recorded. This investigation lasted from February to September, 1954, and its results are given in *Figure 16* and Appendix 19.

Results

(a) The average temperature of jerboas on wet diet is 37° C (36·9° C for males and 37·1° C for females).

(b) The average body temperature of jerboas on prolonged dry diet, for more than six months, is slightly lower, at 36·8° C (36·7° C for males and 36·9° C for females). These differences are not significant but it does seem that one of the chronic effects of dry diet is a lowering of the body temperature.

(c) The average body temperature of white rats is 37·5° C; it is the same for both sexes.

The differences in the body temperatures of the rat and the jerboa are significant and their interpretation appears in the discussion that follows.

ZONE OF BODY-TEMPERATURE REGULATION IN JERBOAS ON WET AND DRY DIETS

Having determined the jerboa's body temperature, we next examined the extent to which the animal can maintain its temperature constant under external temperatures from $-2°$ C to 35° C. To establish this zone of regulation, we conducted the following series of experiments.

Experiments

(a) The method of taking body temperature described in Appendix 1 was used. The air-conditioning chest described in Appendix 1 and special cages prepared for the purpose served as equipment.

(b) Three wet-diet and two dry-diet jerboas were used for the investigation.

(c) The animals were adapted to different chest temperatures.

(d) The duration of the experiment was three hours, and the experiment was repeated the next day under identical conditions.

(e) Early in the morning, food was removed from the cages and the experiments were conducted in the afternoon of the same day.

(f) After the air-conditioning chest was brought to the desired temperature, the animals were introduced into their cages. Three hours later they were removed one by one and their rectal temperatures taken immediately.

This series of experiments was started on January 26 and completed on February 17, 1955.

Results

The results of this series of experiments on the zone of body-temperature regulation of jerboas on wet and dry diet are presented in *Figure 17* and Appendixes 9 and 10. Our findings indicate the following:

1. Jerboas on dry diet have a lower body temperature than those on wet diet.

2. Both wet-diet and dry-diet jerboas maintain their body temperature constant at ambient temperatures between 5° C and 20° C.

3. Between 5° C and 0° C both groups of animals show a rise in body temperature.

4. Below 0° C both groups show a slight drop in body temperature.

5. Jerboas on wet diet at an external temperature of 25° C manifest a slight drop in body temperature, more marked at 30° C. At 35° C a rise in body temperature takes place, exceeding the normal temperature by 0·5° C.

6. Jerboas on dry diet show a slight increase in body temperature, starting with the ambient temperature of 25° C. From 25° C to 35° C the rise is accentuated, reaching 38° C at the ambient temperature of 35° C.

There is every indication that the temperature of thermal neutrality is 30° C for wet-diet jerboas and close to 25° C for dry-diet jerboas.

7. For both wet-diet and dry-diet jerboas the extent of body temperature variations, for the whole range of environmental temperatures from −2° C to 35° C, differs; it is of the order of 0·5° C for jerboas on wet diet and 1·5° C for jerboas on dry diet.

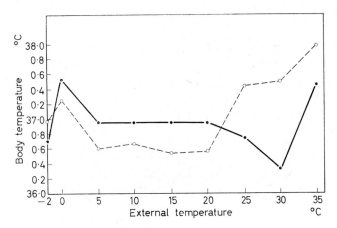

Figure 17. Zone of body-temperature regulation of jerboas on wet (———) and dry (-----) diets.

Interpretation

Jerboas on dry diet are able to regulate their body temperature at a lower level than jerboas on wet diet at external temperatures below 20° C; however, above 20° C, where evaporative water loss begins to play an important part in thermoregulation, their body temperature rises more rapidly, as the duration of three hours at an elevated temperature strains their water reserve.

COMPARISON OF ZONES OF BODY-TEMPERATURE REGULATION IN JERBOAS AND RATS

The above-mentioned results led us to undertake a comparative study of the thermoregulation of jerboas and rats under identical environmental conditions.

Experiment

Three jerboas and three rats were studied during the period from June 6 to October 1, 1955, at external temperatures between −2° C and 40° C.

The same experimental procedure as that described in the previous section was adopted.

The diet was the same (i.e. humid) and the conditions of testing were identical for all animals. They were submitted to a period of conditioning before starting the final tests.

For the external temperatures of 30° C and 35° C the experiments were repeated in two different months for greater accuracy, and the test for 40° C was added to establish the upper limits of heat tolerance. Unfortunately, tests for environmental temperatures below −2° C, to establish the lower limits of cold tolerance, could not be performed owing to lack of adequate equipment.

Results

The results of the comparative study on the zone of body-temperature regulation of jerboas and rats are presented in *Figure 18* and Appendix 10. These findings indicate the following:

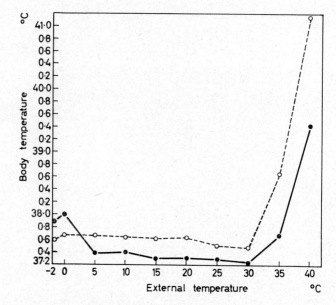

Figure 18. Zone of body-temperature regulation of jerboas (———) and white rats (-----).

1. The jerboa's body temperature remains on a lower level than the rat's at all environmental temperatures from 5° C to 40° C.

2. The rat maintains its temperature constant (about 37·5° C) at ambient temperatures from −2° C to 30° C; but at higher temperatures the animal is subject to stress and its body temperature increases by 1·2° C when it is kept for three hours in the enclosure at 35° C, and by 3·6° C when the temperature is raised to 40° C.

3. For the same range of temperature, we have seen that the jerboa reacts differently, and its body temperature is less stable between 30° C and

CONCLUSIONS

−2° C; however, after three hours at 35° C the animal's temperature rises by only 0·5° C, and after three hours at 40° C by only 2·2° C. This shows that the jerboa's thermoregulation is more efficient than the rat's for ambient temperatures above thermal neutrality. This fact is worthy of further comment.

It is interesting to note that observation of the behaviour of these two different species of animals partly explains the facts. At 35° C the jerboa sleeps profoundly, whereas the rat is agitated. At 40° C the jerboa is still in a state of deep sleep and secretes saliva profusely. The absence of muscular work and the evaporation of water by salivation are two favourable factors in the fight against heat.

Discussion of results

The thermoregulation of homotherms has been widely investigated in many countries. Gelineo (1934) has demonstrated that the range within the limits of which body temperature is maintained can be modified through acclimatization (experimental conditioning). Giaja (1938) studied the phenomenon of hypothermia in homotherms and reviewed the subject of thermoregulation.

The capacity of an animal to regulate its internal temperature in order to withstand the exigencies of the environmental heat stress has, no doubt, an evolutionary significance. This subject has been adequately reviewed by Prosser (1952) and recently by Hart (1957) and Fry (1958).

Thus, a rodent like the jerboa, which has in the course of centuries adapted itself to a desert environment by gradually modifying its behaviour and developing physiological and genetic adaptations, is expected to react better to elevated temperatures.

CONCLUSIONS

1. The jerboa's body temperature is 37° C (36·9° C for the male and 37·1° C for the female).

2. Jerboas fed on a dry diet for more than six months have a slightly lower body temperature than those fed on a wet diet. They behave as though their energy expenditure were reduced. This matter will be re-examined in the next chapter.

3. The rat's body temperature (37·5° C) is 0·5° C higher than the jerboa's (37° C).

4. The jerboa's body temperature is maintained on a lower level than the rat's at external temperatures between 5° C and 40° C.

5. The thermal neutrality of both the jerboa and the rat appears close to the environmental temperature of 30° C, since in the neighbourhood of this ambient temperature the jerboa's body temperature is 37·2° C and the rat's 37·5° C.

6. At temperatures between 30° C and 5° C both species show a similar behaviour.

7. At temperatures above 30° C, however, a fundamental difference is observed between the jerboa and the rat. After remaining for three hours at the external temperature of 35° C the rise in the rat's body temperature is

much more marked than the jerboa's (1·2° C against 0·5° C). The jerboa sleeps peacefully, but the rat becomes agitated. At the environmental temperature of 40° C the differences of behaviour are accentuated, as the temperature of the rat rises by 3·6° C, whereas that of the jerboa by only 2·2° C. The jerboa is still in a state of sleep, with profuse secretion of saliva, whereas the rat's mechanisms of thermoregulation are forced. The jerboa, therefore, presents an adaptation of its thermoregulation to elevated temperatures that must have been established through an evolutionary process.

8. Jerboas on dry diet, compared with those on wet diet, present certain characteristics at ambient temperatures from 5° C to 20° C; their body temperature is slightly lower. But at ambient temperatures from 20° C to 35° C their body temperature rises above that of jerboas on wet diet and their ability to regulate their body temperature is less efficacious. They become heat sensitive, as the heat loss through evaporation of water is limited by the absence of water reserves.

9. A study of these results indicates that the jerboa is physiologically and, perhaps, genetically endowed with specific thermoregulatory mechanisms which enable it to sustain elevated external temperatures. It is an animal adapted to the desert and its behaviour differs profoundly from that of the Wistar rat.

8
ENERGY METABOLISM
IN RELATION TO EXTERNAL TEMPERATURES

The energy metabolism of the jerboa (*Dipus aegyptius*) has not previously been investigated. The only studies made recently were by Haberey and Kayser (1959) and Haberey (1960) on the thermogenesis in the cold of certain species of rodent of Saharan origin, closely related to the species we have studied. Hence we had to answer the following questions:

What is the zone of thermal neutrality of the jerboa compared with that of the rat?

What is the rate of the jerboa's basal metabolism?

How do the total oxidative processes of the jerboa vary in relation to external temperature and in what respect are they different from those of the rat?

What is the repercussion of the ingestion of dry diet instead of wet diet on the curve of respiratory exchanges in relation to temperature?

What are, according to the responses of the two species, the limits of resistance to high temperatures?

Before describing our experiments it will be useful to refer briefly to certain classical findings on the basal metabolism and on thermoregulation.

BASAL METABOLISM

The basal metabolism represents the oxidation that takes place when the animal is at complete rest in an ambient temperature equivalent to its thermal neutrality, having fasted a sufficient time to allow the extra heat resulting from the specific dynamic action (SDA) of food to be dissipated. These oxidative processes represent especially:

1. Oxidation connected with the maintenance of the constitution of cells in the entire organism. All the constituents, with the exception of desoxyribonucleic acid, are in constant renewal. In the adult organism the syntheses balance the degradations but the rapidity of renewal varies with the types of cell and the constituents considered. This renewal or synthesis of all the cellular materials consumes energy, which is derived from oxidative processes. These processes constitute what is called the fundamental or elementary respiration of the tissues. They form the subject of numerous researches by modern biochemists, and it is known that the chemical energy liberated is transferred to the phosphoric bonds named 'rich in energy' in order to serve all cellular works: syntheses, mechanical, osmotic and electrical functions, etc. In the homotherm they are of greater intensity than in the poikilotherm, taking into account the temperature coefficient of Van't Hoff's rule (Le Breton, 1926).

2. At the level of the basal metabolism there are oxidative processes connected with the various muscular works which Lefèvre (1911) called the

'physiological service': contraction of the heart, respiratory muscles, tonus of various skeletal muscles, etc.

3. When an animal is placed at the temperature of its thermal neutrality, the value of the basal metabolism measured is related to the temperature of its natural environment. This is the phenomenon of thermal adaptation studied extensively in the rat by Gelineo (1934), and subsequently the subject of several publications.

The closer the temperature in which the animal lives is to that of its thermal neutrality the lower is the basal metabolism, because under the influence of cold oxidation increases by the secretion of hormones in order to compensate and equalize the thermolysis. Adrenalinemia rapidly reverts to normal when the animal is placed at the temperature of thermal neutrality for the measurement of basal metabolism; this is not the case in thyroxinemia and this is why the rat must be placed at a temperature of about 25–28° C for 15–20 days in order to obtain the minimum level of oxidation.

These are the considerations on thermal adaptation which led us, for our comparison between the jerboa and rat, to allow the animals of the two species to live in the same enclosure and at the same temperature; since the thermal neutrality temperatures of the two species are close the comparisons would, because of this, retain their true value. Moreover, the excitation of the peripheral thermosensitive nerve terminals was suppressed to the maximum, as the temperature of the laboratory unit was 25° C.

To what unit should we refer the values found? This is not the place to dwell on the ancient and endless discussions concerning the law of body size versus the law of surface area. It is obvious that when one wants to study the action of a dietary factor (SDA), or of a pharmacological one, or of the ablation of a gland of internal secretion, the animal itself can serve as a control in measuring the respiratory exchanges before and after the operation. But when the question arises of comparing two animals, whether they belong to the same species or to different species, a unit of reference is indispensable, and we have adopted the kilogram of live weight in the same way as other authors have (following the discussions on the significance to be attributed to the geometric surface of the body). The unit of body weight is even more justified here since the two animals are rodents of similar size.

THERMOREGULATION

We do not propose to present the history of this vast subject, on which a number of publications have recently appeared, notably on the problem of hibernation (Kayser, 1940). We shall confine our comments to certain classical findings that will help to define our experimentation. The sum of the mechanisms which enable an animal homotherm to maintain its body temperature constant in spite of important variations in external temperature constitutes thermoregulation. It is a type of unconscious neurovegetative regulation.

We denote the nervous centres and the two great systems of reflex innervation: the one of physical regulation that intervenes in thermolysis and the other of chemical regulation that regulates heat production by increasing the oxidation rate.

Physical regulation

For the physical regulation the nervous command is direct. It intervenes by modifying the size and quality of active cutaneous surface, the attitude, the peripheral vasodilatation and vasoconstriction, the cutaneous water loss, the pulmonary ventilation, the pilomotor reflexes, etc. All these factors which regulate the extent of thermolysis vary in negative proportion to the amount by which the external temperature is above or below the animal's point of thermal neutrality. The important function of these several mechanisms of thermolysis has been investigated in different species under varying conditions of temperature, relative humidity, renewal of air at body surface, etc.

Among the physical factors of importance are, according to environmental temperature, the separate parts taken in thermolysis by the latent heat of water evaporation, by radiation from the surface and by convection and conduction. We shall see, in the discussion of the results, how the loss of latent heat evolves with increase of ambient temperature and the part played, according to species, by sweating, salivation and thermal polypnea at the point where heat loss by radiation is annulled. It is evident that if the external temperature is above that of the skin the subject will store heat energy, and it is then that the latent heat of evaporation intervenes in order to permit the dissipation not only of heat energy from oxidation but also of that from 'overheating'.

Chemical regulation

The chemical regulation does not intervene except in the zone of cold since, by definition, the level of basal metabolism is the lowest level of oxidation. As soon as the external temperature is lowered, not only the physical regulation intervenes to limit the loss but also the chemical regulation is able to adjust the body's thermogenesis according to the increasing rate of thermolysis.

The true chemical regulation, which is the increase of oxidation through hypersecretion of hormones (thyroidal hormones, adrenalin and noradrenalin, intermedine), should be differentiated from the increase of oxidation through shivering. In the small animal it is the hormonal secretions that play the principal role, whereas in the dog, for example, and in Man, shivering begins before this (Chatonnet, 1961). For the role and mechanism of hormonal action in cold we refer the reader to the complete publication of Thibault (1949) on the rat. It is to be noted that adrenalin acts by increasing glycogenolysis but not by the oxidation of sugars, and, as shown by Schaeffer and Polak (1938), it is the oxidation of fatty acids that is catalysed by this hormone. With regard to thyroxine, it increases oxidation by a direct effect, as it also permits a maximum action of adrenalin (potentization).

EXPERIMENTS

We shall now explain the comparative studies made by us on the energy metabolism of the jerboa and rat, with all animals submitted to the same experimental conditions. The method of indirect calorimetry described in

Appendix 1 was used. Each experiment comprised the simultaneous studies of a jerboa on dry diet, a jerboa on wet diet and a rat on wet diet under identical conditions. The same animals were used, as far as possible, for the whole range of ambient temperatures from 0° C to 45° C. Furthermore, to take into account individual variations, several series of animals were subjected to the same type of experiment.

RESULTS

Figure 19 gives the curves established on the basis of mean values for all the animals subjected to experiments. The experimental findings are presented in Appendix 11. Examination of these results reveals the following findings.

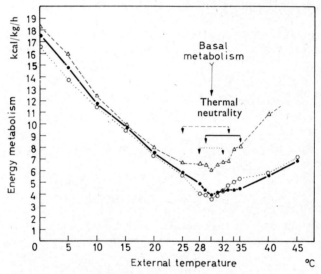

Figure 19. Energy metabolism in relation to external temperature of jerboas on wet diet (———), dry diet (.....) and rats on wet diet (-----).

Basal metabolism

It is observed that the temperature at which the jerboa's respiratory exchanges are at a minimum is 30° C. The rat's minimum exchanges also take place at 30° C, as shown by classical researches undertaken prior to ours.

The basal metabolism of the jerboa on wet diet is 3·98 kcal/kg/h and that of one on dry diet 3·65 kcal/kg/h. But the white rat (Wistar) on the same wet diet shows a basal metabolism of 6·16 kcal/kg/h. Therefore, although the jerboa has a smaller size than the rat (160 g instead of 250 g) its oxidative processes are lower.

Thermal neutrality (heat loss equal to minimum heat production).

The thermal neutrality of a jerboa on dry diet is close to 30° C (between 27° C and 32° C), as is apparent from the curve obtained from mean values.

RESULTS

The curve of a jerboa on wet diet, however, presents almost a plateau at the zone of thermal neutrality between 29° C and 35° C (4·33 and 4·56 kcal).

The interesting fact is that rats subjected to the same wet diet and having lived at 25° C in an atmosphere equally humid, present a plateau even more marked on the curve of respiratory exchanges between 25° C and 33° C (6·77 and 6·90 kcal); in other words, it shows a 'zone' of thermal neutrality instead of the 'point' generally observed. Should we attribute this to the conditions under which the rat lived or to failure of our technique to detect the light variations of metabolism?

We propose later to undertake the study of this phenomenon by using a very precise method that makes possible measurements of short duration by eliminating movement and by modifying the conditions of life.

Heat production

For the range of external temperatures from 0° C to 25° C, the level of the jerboa's heat production is somewhat lower than that of the white rat. Also, the dry-diet jerboa shows a slightly lower level of heat production than the wet-diet jerboa. At 0° C, the rat produces 18·34 kcal/kg/h, whereas the wet-diet jerboa produces 17·56 kcal/kg/h and the dry-diet jerboa 16·59 kcal/kg/h.

The heat-production levels are of almost the same order and remain parallel between 0° C and 20° C, since at 20° C the rat produces 8·02 kcal/kg/h, the wet-diet jerboa 7·50 kcal/kg/h and the dry-diet jerboa 7·48 kcal/kg/h. But from more elevated temperatures up to the level of thermal neutrality the heat production of the two species diverges until at 30° C, as we have seen, it is 6·16 kcal/kg/h for the rat, 3·98 kcal/kg/h for the wet-diet jerboa and 3·65 kcal/kg/h for the dry-diet jerboa.

Influence of temperatures above thermal neutrality on respiratory exchanges (zone of physical regulation)

Beyond the zone of thermal neutrality, at higher ambient temperatures, the differences of oxidation level are accentuated and indicate a very different behaviour for the two species.

At the environmental temperature of 40° C, the levels of calorification in kcal/kg/h are, for the rat, 10·89, for the jerboa on wet diet, 5·60, and for the jerboa on dry diet, 5·83. We see that the difference of oxidation level that existed between the rat and jerboa at the temperature of 30° C—namely, at the level of basal metabolism—is greatly increased. Whereas the rat augments its oxidative processes by 76 per cent (6·16 to 10·9 kcal/kg/h) when the ambient temperature rises from 30° to 40° C, the dry-diet jerboa augments by 60 per cent (3·65 to 5·83) and the wet-diet jerboa by only 40 per cent (3·98 to 5·59). The temperature of 40° C is the upper limit that the rat can support during one hour.

Moreover, the jerboa can tolerate an ambient temperature of 45° C for about one hour, whereas the rat is unable to do so. In fact, at the temperature of 45° C the calorification of a jerboa on wet diet is only 6·91 kcal/kg/h, or an increase of 70 per cent over the basal metabolic rate, and that of a jerboa on dry diet is 7·18 kcal/kg/h, or an increase of 97 per cent. These higher temperatures are supported by jerboas by augmenting their heat dissipation by evaporation through the aid of salivation and wetting of body parts. At this point the difference in behaviour between jerboas on wet diet and those on dry diet is marked.

Discussion of results

The above-mentioned findings indicate that the zones of thermal neutrality and the basal metabolic rates of rats and jerboas on wet diet and of jerboas on dry diet differ.

Basal oxidation: It is important to emphasize that the oxidative processes of the rat, 'placed' at the level of basal metabolism, are greater than those of the jerboa, even after adaptation to 25° C. Our figures are higher than those obtained by other investigators, perhaps because of the animal's breed, but the lowest values published for the rat are of the order of 4·5–5 kcal/kg/h (Thibault, 1949; Brody, 1945; Spector, 1956) and therefore higher than the jerboa's oxidation rate of 3·98 and 3·65 kcal/kg/h on wet and dry diet respectively. We also emphasize the fact that, in spite of its small size, the jerboa has a lower heat production, which seems to be the result of its adaptation to an existence of under-nutrition for several millennia, having been led gradually to restrict its 'surplus' energy expenses. Life in the desert requires, in fact, not only an adaptation to the elevated and very variable temperatures in the course of the daily 24 hours, but also, because of the nature of desert flora and fauna, a limitation of the animal's daily intake in calories and water.

Nevertheless, we think that the factors originally responsible for the characteristics of the jerboa's metabolism must be the association of high temperature and inanition of water. This scarcity of water, as we have seen, entails the development, in the course of evolution, of a type of urinary secretion whose study we propose to undertake in the near future. Every ingestion of increasing quantities of food entails an increased elimination of water through the feces, and particularly through the kidneys, if the percentage of proteins is augmented. The ingestion of dry diet, represented by cereals, involves an intake of glucosides, whose oxidation is a better source of water per calories liberated than the oxidation of lipids or proteins (here urea must be eliminated). We have also observed that the basal metabolic rate is slightly lowered if the jerboa is kept on dry diet.

There is no need here to describe in greater detail the calculations for our experiments on the heat balance and water balance of the three sets of animals—jerboas and rats on wet diet and jerboas on dry diet—during a period of six days. *Tables 10* and *11*, in Chapter 10, indicate that at the temperature of 25° C the jerboa on wet diet stores less water and calories (in the form of fat) than the rat. When the intake of water is very limited, the jerboa manages to balance its losses with the water derived from its oxidative processes.

Behaviour in the zones of low and high temperatures: When the temperature is lowered, the jerboa behaves in a normal manner; that is, its exchanges augment along a straight line with the lowering of external temperature. The slope of this line corresponds to an increase of oxidation by 0·4 kcal/°C whether the diet is wet or dry. The calorification levels are parallel, the values of heat production always being lower for the jerboa on dry diet.

The oxidative processes that supplement the basal metabolism and those that take place in the fight against cold, thermoregulatory oxidation, are almost the same whether the jerboa is on wet diet or on dry diet, namely:

$16 \cdot 59 - 3 \cdot 65 = 12 \cdot 94$ kcal/kg/h for jerboas on dry diet
$17 \cdot 56 - 3 \cdot 98 = 13 \cdot 58$ kcal/kg/h for jerboas on wet diet

Here it is a matter of adjusting thermogenesis to thermolysis, and thermolysis is commanded by the surface of heat dissipation, which is constant (no matter what the diet is), and by the external conditions, which remain the same. Hence, an identical increase of oxidation results in both cases. It is noteworthy that in this zone of cold the oxidative processes arising from thermoregulation are more important per kg/h in the jerboa than in the rat—13 kcal/kg/h for a drop of 30° C in external temperature against 12 kcal/kg/h in the rat for the same diminution: here the jerboa's active surface of heat dissipation is proportionally greater than the rat's, and this commands increased heat production.

In the fight against heat, which is carried on by the evaporation of greater quantities of water beyond 30° C and solely by this mechanism in external temperatures higher than 36° C, the jerboa is more successful than the rat, and a jerboa on wet diet is even more efficient. The causes of the heat tolerance of this desert rodent can be summarized as follows:

1. The jerboa's basal metabolic rate is lower—namely, 3·65 and 3·98 kcal/kg/h instead of 6·16 kcal/kg/h for the rat. Therefore, per 100 g of body weight, it has to dissipate less heat and borrow less water from its reserves, which we have experimentally established to be greater than those of the rat.

2. The elevation of external temperature in the rat provokes a state of agitation and movements that are a source of complementary heat dissipation, whereas in the jerboa it instigates a state of relaxation, a diminution of muscular tonus and a state of lethargy, so long as it is not 'forced'. This reflex adaptation is one of the principal causes of the jerboa's resistance to heat, and it could be said that a set of conditioned 'estivation' reflexes are established similar to what is realized in 'hibernation' for adaptation to cold.

3. It must also be noted that the animal's posture and the phenomenon of reflex salivation contribute to its adaptation to heat, contrarily to the rat, which pants (polypnea). A jerboa on wet diet can support an environmental temperature of 45° C for more than one or two hours before manifesting a body temperature which causes irreversible impairment. Dry diet is associated with economy of water, but as the reserves are limited resistance to higher temperatures is low.

In our opinion this discussion summarizes the principal conclusions of this chapter—fuller explanations of our findings follow.

9
INSENSIBLE PERSPIRATION AND EVAPORATIVE WATER LOSS IN RELATION TO THERMAL ENVIRONMENT

INTRODUCTION

At the beginning of the previous chapter we explained how the heat of oxidation is dissipated in two ways:

(a) *Heat loss from body surface (radiation, convection, conduction):* The surface where thermolysis occurs is the active surface defined by Bohnenkamp and Pasquay (1931). It is conditioned by the animal's attitude and varies according to the 'heat dissipating' qualities of the species. The quality of the surface, fur, feathers, etc., all affect the size of the 'zone of thermal neutrality'.

(b) *Heat loss through the latent heat of vaporization of water:* The quantity of heat lost per gram of water evaporated is related to temperature. According to Robinson (1949), one gram of water vaporized at 33° C, which represents approximately the temperature of the skin, removes 0·58 kcal from the body. At environmental temperatures above skin temperature, evaporation is the only way to lose heat in order to assure maintenance of the body temperature. By this mechanism the animal loses, besides the heat liberated by its metabolic reactions, the heat gained from the environment.

Hence there are two ways of heat dissipation by vaporization of water: (a) insensible perspiration and (b) evaporation of water, by sweating if the animal possesses sweat glands, by thermal polypnea (panting), or by salivation if saliva is spread on the body surface. Consequently, heat tolerance is closely related to the animal's capacity to vaporize water in sufficient quantity.

In a hot environment, the problem of thermoregulation and thermal equilibrium is related to the possibilities of water loss in an efficacious form (vaporization) and in quantities that do not upset the balance of physiological functions.

During recent years the mechanisms of heat dissipation in Man and in animals have received considerable attention. We shall not dwell upon the classical publications of Dill (1938), Adolph (1947), Kuno (1934) and many others. We simply mention that Adolph drew attention to the differences in behaviour of species in this regard. He found that thermolysis by polypnea is considerable in dogs and cats, but limited in guinea-pigs, rabbits and small rodents.

Salivation is a process limited to some species, and as stated above, it is observed in the jerboa. Herrington (1940) reported that the mouse, rat and guinea-pig use salivation as an emergency mechanism for heat regulation. Prouty (1949) found that the cat exposed to elevated temperatures spreads saliva on its body parts, thus aiding thermolysis when its body temperature tends to rise. Schmidt-Nielsen (1954) observed that when the kangaroo rat's body temperature rises to about 42° C a profuse secretion of saliva occurs which, on evaporating, prevents further rise of temperature.

It is also necessary to take into consideration certain data on desert rodents. The Schmidt-Nielsens (1950) observed in desert rodents, hamsters and wild mice a total evaporation of about 0·5 mg of water per millilitre of oxygen utilized (12 per cent of the thermolysis). The same authors (1949) found that white rats evaporate twice as much water as desert rats (of the *Heteromyidae*) per millilitre of oxygen used (25 per cent of the thermolysis).

Schmidt-Nielsen (1949) found that heteromyids breathing dry air lose 0·45 mg of water per millilitre of oxygen used. As the metabolic water they derive from the oxidation of food amounts to 0·65 mg per millilitre of oxygen their water balance is positive, by this fact alone, in dry air. The same author (1954) reported that kangaroo rats show a low evaporation from the lungs, as the air expired is saturated at their nose temperature, which is 10° C lower than that of the body. White rats also show low evaporation from the lungs, but their total evaporation from other parts of the body is higher than that of the kangaroo rat.

Thermolysis through latent heat of evaporation in ambient temperatures equal to or lower than thermal neutrality

In this temperature range only insensible perspiration occurs, diminishing gradually as the external temperature drops and oxidative processes increase (Mayer and Nichita, on the rabbit, 1929).

At the level of basal metabolism the percentage of thermolysis in the form of latent heat varies according to species. The classical figures for Man are from 22 to 35 per cent. Bazett (1949) indicated 24 per cent at the level of the skin and 10 per cent for pulmonary insensible perspiration. Measurements have been made by Kayser (1930) and Kayser and Dontcheff (1941) on the pigeon for thermal neutrality and higher temperatures.

Newburgh and Johnson (1942) have reviewed the literature on insensible perspiration. Among the most recent studies are those of the Schmidt-Nielsens (1952) on desert mammals and of Kayser (1954) on the general question of water metabolism and thermoregulation. We shall now describe our personal experiments.

EXPERIMENTS

In order to define the role of heat loss by evaporation in the thermolysis of jerboas and white rats, the following questions had to be answered:

1. What is the evaporative water loss of the jerboa under environmental temperatures of 0–45° C?
2. What is the effect of dry diet on the evaporative water loss of the jerboa?
3. What is the evaporative water loss of the rat, compared with the jerboa, under environmental temperatures of 0–45° C?

For this purpose a series of controlled experiments was undertaken simultaneously on wet-diet jerboas, dry-diet jerboas and white rats. The conduct of these experiments and the technique used are described in detail in Appendix 1 under the title 'Insensible perspiration' (see p. 107).

Results of the experiments appear in Appendix 12, and graphically in *Figures 20–22*.

RESULTS

Insensible perspiration and heat loss by evaporation of water

As the basal metabolism is normally measured at 30° C, we studied the insensible perspiration (cutaneous and pulmonary water vaporization) at the same temperature, of the rat, jerboa on wet diet and jerboa on dry diet. We have found that the rat loses 1·136 g/kg/h, the jerboa on wet diet 0·714 g/kg/h and the jerboa on dry diet 0·659 g/kg/h.

At external temperatures of 30–0° C, insensible perspiration rises gradually until at 0° C it reaches 2·749 g/kg/h (141 per cent) for the rat, 1·201 g/kg/h (68 per cent) for the jerboa on wet diet and 0·949 (43 per cent) for the jerboa on dry diet.

At higher ambient temperatures, above 30° C, the level of heat loss by latent heat presents great divergences in the three groups of animals, as shown in *Table 7* and *Figure 20*.

Table 7

Insensible perspiration of jerboas and rats

External temp.	Specimen	Water loss (g/kg/h)	Mechanism
32° C	Rat (wet diet)	2·139	Slight salivation
	Jerboa (wet diet)	1·316	No salivation, asleep
	Jerboa (dry diet)	0·776	No salivation, asleep
35° C	Rat (wet diet)	3·033	Salivation and polypnea
	Jerboa (wet diet)	1·502	Slight salivation, asleep
	Jerboa (dry diet)	1·371	Slight salivation, asleep
40° C	Rat (wet diet)	14·743	Limit of tolerance
	Jerboa (wet diet)	8·268	Profuse salivation, wetting of body parts
	Jerboa (dry diet)	5·061	Profuse salivation, wetting of body parts
45° C	Rat (lethal temp.)		
	Jerboa (wet diet)	14·217	Profuse salivation and wetting of body parts
	Jerboa (dry diet)	13·798	Profuse salivation and wetting of body parts

The part of latent heat in thermolysis

The values of the latent heat of insensible perspiration, expressed in percentage of the total heat produced in kcal/kg/h at environmental temperatures of 0–45° C, are presented in *Figure 21* and Appendix 12 (we have adopted as the mean value 0·59 kcal per gram of water vaporized). Our findings indicate the following:

The minimum percentage of thermolysis by latent heat (cutaneous and pulmonary evaporation) occurs at the environmental temperature of 0° C, at which point the evaporative heat loss of the rat is 9·1 per cent, the jerboa on wet diet 4·6 per cent, and the jerboa on dry diet 3·7 per cent.

At environmental temperatures of 0–30° C this percentage rises very slightly, until at a temperature of 30° C the latent heat loss of the rat reaches 11·4 per cent, the jerboa on wet diet 10·2 per cent and the jerboa on dry diet 8·7 per cent.

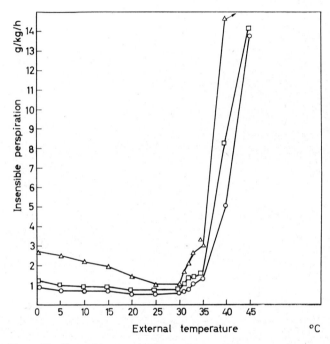

Figure 20. Insensible perspiration and evaporative water loss in relation to thermal environment.
△ rat on wet diet, □ jerboa on wet diet, ○ jerboa on dry diet

At the environmental temperature of 35° C the latent heat loss reaches 23·1 per cent for the rat, 17·2 per cent for the jerboa on wet diet and 17·09 per cent for the jerboa on dry diet.

At the environmental temperature of 40° C the latent heat loss of the rat reaches 90 per cent, admitting that there is no profuse saliva in the water measured, that of jerboas on wet diet 87 per cent and that of jerboas on dry diet 54 per cent.

Environmental temperatures above 40° C are lethal to the rat, whereas jerboas on wet diet and on dry diet continue to resist these high temperatures up to 45° C by salivating profusely and wetting certain parts of their body. Their thermolysis by latent heat loss cannot be calculated here, as part of the water is not vaporized and consequently the figures of 121 per cent and 113 per cent are recorded (see Appendix 12).

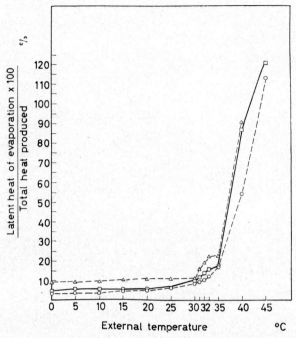

Figure 21. Latent heat of evaporation in relation to total heat produced.

△ rats on wet diet, □ jerboas on wet diet, ○ jerboas on dry diet

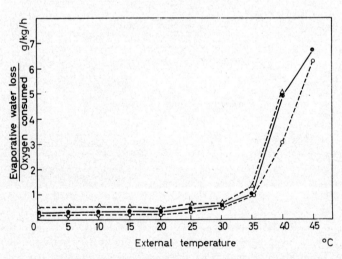

Figure 22. Evaporative water loss in relation to oxygen consumed.

△ rats on wet diet, ● jerboas on wet diet, ○ jerboas on dry diet

RESULTS

Ratio of evaporative water loss to oxygen consumed

Appendix 12 gives the ratios of evaporative water loss to oxygen consumed, water and oxygen being expressed in g/kg/h. *Figure 22* illustrates these results. The data indicate:

For environmental temperatures of 0–45° C a parallel exists between the H_2O/O_2 ratio and the percentage of thermolysis by latent heat in the rat and in jerboas on wet and dry diet.

This ratio increases slightly but steadily between 0° C and 30° C; in the rat it passes from 0·516 to 0·652, in the jerboa on wet diet from 0·259 to 0·576 and in the jerboa on dry diet from 0·207 to 0·490.

Above the temperature of 30° C, the H_2O/O_2 ratio rises sharply, and at 35° C reaches 1·3 in the rat, 0·96 in jerboas on wet diet and 0·96 in jerboas on dry diet. At 40° C, these ratios are 5·04 in the rat (which constitutes the limit of its tolerance), 4·94 in jerboas on wet diet and 3·09 in jerboas on dry diet.

At the ambient temperature of 45° C, the H_2O/O_2 ratio in jerboas on wet diet is 6·725 and in those on dry diet 6·297. However, as we have pointed out, at this temperature water loses its univocal significance and consists of vaporized water (latent heat and profuse water).

INTERPRETATION OF RESULTS

For external temperatures from 0° C to 45° C the share of latent heat (vaporization of water) in thermolysis augments at first progressively and then abruptly beyond the thermal neutrality. In absolute values the vaporization of water corresponds to the insensible perspiration at its minimum, at the temperature of thermal neutrality, when the oxidative processes themselves reach their minimum level. But at ambient temperatures from 30° C to 0° C, evaporative water loss increases only slightly, whereas the calorification triples and quadruples. This occurs because calorification is in direct relation to the increase of thermolysis, and, consequently, there is need for greater intake of oxygen and greater intensity of the phenomena of respiratory function. The inspired air must be heated and humidified within the limits of body temperature. Water loss through the lungs at 0° C cannot but be higher than what it is at the ambient temperature of 30° C.

On the other hand, the evaporative water loss at the level of the skin must diminish as external temperatures decrease. This diminution, however, is inferior to the pulmonary increase, so that the overall evaporative water loss at 0° C is greater than what it was at 30° C.

The method that we devised to estimate the quantity of water lost through evaporation is described in Appendix 1. It has the following advantages:

1. The weight of CO_2 increment over O_2, as calorification rises owing to environmental temperature, is duly accounted for.
2. The fact that the animals lived in confined air and that consequently the humidity of the air increased during the experiment does not change the comparative values of the results. It is *a priori* true that at higher humidities evaporative water loss is suppressed because of saturation; but this suppression is a common factor, and, in fact, by the use of confined air the

experimental conditions are brought nearer to those existing in the natural humid habitat of jerboas. It may be argued that the utilization of dry, circulating air has its advantages, but then the results would not be more representative of what in reality takes place, and our aim is to understand the natural mechanisms of the jerboa's adaptation. Moreover, the method chosen is accurate to one milligram of water, thus giving results of valid interpretation.

We are of the opinion that for these reasons thermolysis by the latent heat of evaporation, expressed in percentage of total heat produced (or lost), shows values lower than those reported in the literature on experiments conducted in dry, circulating air. By the latter method the latent heat of evaporation is considered to be 25 per cent at thermal neutrality, whereas according to the method adopted by the author the results obtained at the ambient temperature of 30° C are 11·4 per cent for the rat, 10·2 per cent for jerboas on wet diet and 8·7 per cent for jerboas on dry diet. The purpose of the investigators who made these measurements was different from ours, and our figures surely describe more accurately that which effectively takes place in the natural ecoclimate of the jerboa.

CONCLUSIONS

The experiments concerning the part played in thermolysis by loss of heat in the form of insensible perspiration or of insensible perspiration plus water of salivation are added to those of the previous chapter to characterize the behaviour of the jerboa from the point of view of thermoregulation.

No matter what the temperature considered is, the part of the heat dissipated by vaporization of water is weaker in this desert rodent than in the rat.

The differences are all the more important and significant for temperatures above the thermal neutrality, rendering evident the adaptation of the thermolysis mechanisms to the conditions of life prevailing in the desert— heat and lack of water. This is well established in summary form by *Tables 8* and *9*.

Table 8

Latent heat loss expressed in percentage of thermolysis

	Environmental temperature				
	0° C (%)	30° C (%)	35° C (%)	40° C (%)	45° C (%)
Rat	9·1	11·4	23·1	90·7	lethal
Jerboa on wet diet	4·6	10·2	17·2	87·2	121·3*
Jerboa on dry diet	3·7	8·7	17·09	54·4	113·3*

* See text for the significance of water at this temperature.

CONCLUSIONS

Table 9

Ratio of evaporative water loss to oxygen consumed (H_2O/O_2)

	Environmental temperature				
	0° C	30° C	35° C	40° C	45° C
Rat	0·516	0·652	1·298	5·045	lethal
Jerboa on wet diet	0·259	0·576	0·965	4·942	6·725*
Jerboa on dry diet	0·207	0·490	0·962	3·089	6·297*

* See text for the significance of water at this temperature.

Besides the differences between the rat and jerboa, these tables show the influence of dry diet on the desert rodent. This diet reduces heat loss by vaporization of water more than the wet diet for equivalent temperatures. It is interesting to note again the two characteristics of the jerboa's behaviour:

1. Inactivity and even sleep in order to diminish the production of heat when temperature rises and to economize water.

2. Salivation to wet parts of the body surface when temperature rises. Is this phenomenon regulated in relation to temperature as are thermal polypnea in the dog and sweating in Man? We have seen at the beginning of this chapter that this mode of thermolysis has been observed in other species, although it has not been sufficiently investigated. We propose to undertake further research with quantitative measurements.

10

THERMOREGULATION OF THE JERBOA AND RAT

The quality that characterizes the homothermic animal is the ability to regulate its production and dissipation of heat so as to maintain its body temperature constant under varying environmental temperatures. The internal milieu that thermoregulation aims to protect is a fluid one. In the living being, water is mobilized for thermal control at all ambient temperatures. Thermal regulation is inconceivable without the adjustment of body fluids. In order to understand the thermoregulation of an animal, therefore, it is necessary to investigate its heat and water balance.

HEAT BALANCE AND WATER BALANCE

In order to discover how two different species, the jerboa and the rat, balance their income and expenditure of energy and water at the ambient temperature of 25° C, we have grouped the results of the experiments described in Chapter 6 and Appendixes 1 and 8. These results are given in *Tables 10* and *11* and graphically presented in *Figures 23* and *24*. Our experimental data indicate the following:

Jerboas on dry diet, compared with jerboas and rats on wet diet, balance income and expenditure of heat and water with maximum economy without

Table 10
Heat balance

Specimen	Income		Expenditure	
		Per 100 g body weight per day (kcal)		Per 100 g body weight per day (kcal)
Rat on wet diet	Wheat	23·0	Metabolism	18·8
	Lettuce	2·9	Stored	7·1
	Total	25·9	Total	25·9
Jerboa on wet diet	Wheat	21·4	Metabolism	16·4
	Lettuce	0·9	Stored	5·9
	Total	22·3	Total	22·3
Jerboa on dry diet	Wheat	13·9	Metabolism	13·9
	Lettuce	0	Stored	0
	Total	13·9	Total	13·9

HEAT BALANCE AND WATER BALANCE

Table 11

Water balance

Specimen	Income		Expenditure	
		Per 100 g body weight per day (g)		Per 100 g body weight per day (g)
Rat on wet diet	Water (natural)	8·3	Feces	1·6
	Water of oxidation	3·6	Urine	4·9
			Evaporation	2·7
			Stored	2·7
	Total	11·9	Total	11·9
Jerboa on wet diet	Water (natural)	3·9	Feces	0·5
	Water of oxidation	3·2	Urine	2·8
			Evaporation	1·7
			Stored	2·1
	Total	7·1	Total	7·1
Jerboa on dry diet	Water (natural)	0·4	Feces	0·3
	Water of oxidation	2·0	Urine	0·6
			Evaporation	1·5
			Stored	0
	Total	2·4	Total	2·4

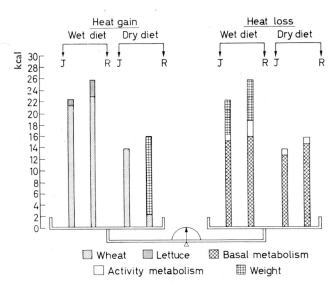

Figure 23. Heat balance of jerboas (J) and rats (R) at ambient temperature of 25° C (per 100 g body weight per day).

being able to store energy and water. The dry-diet jerboa is satisfied with 13·9 kcal and 2·4 g of water per day (for 100 g of body weight), whereas the income of the jerboa on wet diet is 22·3 kcal and 7·1 g of water per day, against 25·9 cal and 11·9 g for the rat on the same diet.

The jerboa and rat on wet diet eat more than they need and store energy and water (increase of body weight).

The jerboa on dry diet makes its maximum economy of water by excreting very little but highly concentrated urine.

The rat on dry diet refuses to eat after the third day and loses weight fast. It derives energy and water from its body stores to keep alive only for a short period (see Figures 23 and 24). As this fasting level of heat and water balance has no comparative value, it was omitted from Tables 10 and 11.

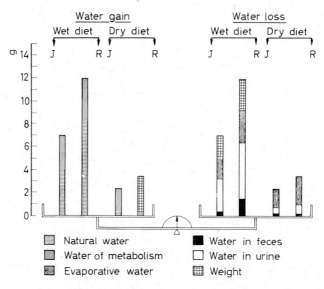

Figure 24. Water balance of jerboas (J) and rats (R) at ambient temperature of 25° C (per 100 g body weight per day).

Interpretation of results

The problem of the animal in the desert is not so much the intensity of heat, which it can avoid by leading an underground and nocturnal life, as the scarcity of water and food. Lack of water drives the animal to economize on its basic needs. Lean eating and little drinking go together. This principle is observed even in the life of human desert inhabitants, the bedouin.

In the natural order of desert factors, scarcity of water leads to rarity of food, and the insufficiency of water and food obliges the animal to eat less. It is this association of scarcity of water and rarity of food which probably forced a desert animal like the jerboa to develop, in the course of its evolution, mechanisms of adaptation that enabled it to live on little food and no drinking water for a long time.

It is evident that in the animal body the economy of energy helps the conservation of water and the two together fortify resistance to high temperatures.

THERMOREGULATION

We have prepared *Figure 25*, which groups distinctly the results of all our researches on the jerboa and rat; namely, the variations in relation to external temperature of (a) body temperature, (b) energy metabolism and (c) insensible perspiration and evaporative water loss.

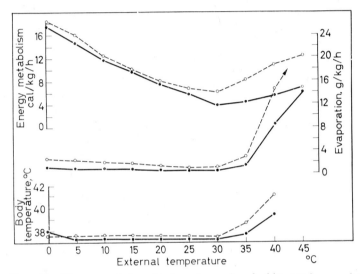

Figure 25. Thermoregulation of jerboas (———) and white rats (-----).

We can see that the levels of the jerboa's body temperature, energy metabolism and insensible perspiration in relation to external temperature are low compared with those of the rat. These differences are maintained at all ambient temperatures from 0° C to 45° C; they are accentuated starting from 25° C and gradually increase in the region of high temperatures, where the jerboa has been able to adapt.

CONCLUSIONS

The scarcity of water, rarity of food and intensity of temperature that characterize the desert environment led the jerboa to reduce its metabolic needs for water and energy, and gradually this forced economy helped to develop mechanisms of adaptation to high temperatures. The rat does not possess these mechanisms, being adapted to a temperate environment. Hence the jerboa is a desert animal *par excellence*, being hardened to the insufficiency of water, deficiency of food and intensity of heat in the desert environment.

11
MAN'S ADAPTATION TO DESERT CLIMATE

CLIMATE AND THE HUMAN RACE

In ancient times Hippocrates wrote on the influence of climate on man, and Galen supported the theory that the inhabitants of temperate zones are mentally superior to those of torrid and cold regions. In the course of centuries, since the observations of Greek writers, many investigators have contributed to the development of knowledge on the effects of climate on man's constitution and behaviour.

In modern times the writings of Jean Bodin, J. Arbuthnot, Montesquieu, L. Febvre, Vidal de la Blache, Ellsworth Huntington, Max Sorre, Clarence A. Mills and others have stimulated interest in the subject. Markham (1947) supported the view that the human race developed in those regions of the world where the environmental temperature averaged 21° C, as life at this temperature does not require extra clothing, special shelter or artificial heat. There are many indications that man's earliest civilizations flourished in temperate regions, that the human organism is better adapted to heat than to cold and that there is no physiological basis for the misconception that the white man is unfit to live in the hot regions of the earth. There seem to be environmental factors other than heat which hindered man's adaptation to tropical and desert climates.

It is known that climatic elements are complex, and their effects on the human organism are equally complex. Moreover, man uses artificial methods of protecting his thermoregulation, such as clothing, shelter, air-conditioning, etc. Thanks to his technology, he can modify climate and survive anywhere on this planet, defying altitude, latitude and adverse climatic conditions.

The question of whether the coloured races are better adapted to hot climates than the white race gave rise to much controversial discussion. It has been reported that coloured people have a more developed network of skin capillaries and a greater number of sweat glands on the skin surface, and that the melanin of their skin is a natural protection against radiation. The role of the fat deposit, steatopygia, of the Kalahari desert bushmen is not known. Moreover, contradictory reports exist on the theory that inhabitants of hot countries show a lower level of basal metabolism.

It has been experimentally established that there are individual differences in matters of sensitivity and resistance to climatic conditions. These differences may be the limiting factors of the capacity for adaptation, as there are individuals in all races who show superior resistance to climatic stress. It has, moreover, been demonstrated that the process of acclimatization is a physiological phenomenon of short duration common to all people who change their climatic environment.

The beneficial effects of the desert climate on health are attributed to the dryness of the air, which facilitates evaporative heat loss, and to the daily range of atmospheric temperature variation, which stimulates the body. According to Sorre (1954), the desert climate is very sanatory. The lack of humidity arrests microbic activity. Wounds heal quickly. Desert aridity is such that even animal carcasses desiccate on the hot soil before decaying. In oases, however, where vegetation and animal life flourish, the atmospheric humidity and stagnant waters favour the development of insect-borne diseases, such as paludism.

According to Sergent (1954), the excessive and dry heat of the desert exerts a depressive action on the nervous system of the white race, provoking digestive disturbances. Infant mortality among the white is very high, and white women are fatigued during the summer season. Conversely, the resistance of the black race to low temperatures in winter is weak.

The continuous migration of populations, the mixture of races and the change of climatic conditions render a study of the effects of climate on man very difficult. The cases observed are few, and the determining causes are multiple. Moreover, hereditary, dietetic, socio-economic and medical factors, intricately related to the subject, have not been sufficiently investigated.

In short, our present knowledge of the climatic influences on human life is inadequate to enable us to establish a climatography of human ecology.

MAN AND THE DESERTS

Man has inhabited the deserts since prehistoric times. Living conditions may have been favourable at times and hostile at others, but on the whole he managed to exist at least on the edge of the deserts.

For example, in the Libyan Desert, where nomadic populations live today, the Tehennu lived many centuries ago as hunters, shepherds and caravan leaders, armed with spears and boomerangs.

According to Weiner (1954), one per cent of the world's population lives in desert climates. All human types are represented among the inhabitants of the ten major desert regions: the Sahara, Australia, Turkestan, Arabia, Argentina, Colorado, Gobi, Kalahari, Thar and Chile. Most of the hot desert regions lie in the northern hemisphere, along the great zone that stretches from the Sahara eastward to the heart of China. This vast area is known as the 'Great Palearctic Desert'.

Pouquet (1951) stated that there are two categories of desert dwellers: (a) the sedentary people of the oases and (b) the nomadic people, pastoral and commercial, who are the true men of the desert, related to sedentary populations.

Among the successful dwellers of the desert are the bedouin of the Orient and the Arabs, Berbers and Tuaregs of Africa. Lhote (1937) described the characteristics of the Arabs, Berbers and Tuaregs as follows: High stature, emaciated features, dry body, knotty members, great physical resistance and sobriety.

Speaking of the bedouin, Montagne (1947) indicated that they have succeeded not only to live in a hostile environment, fighting incessantly against want, thirst, cold and disease, but also to develop their own traditions, customs, mores and laws—a complete 'desert civilization'.

Way of life

In the desert, man's own activities and the essential factors of his environment, such as soil, vegetation, animals and climate, are all conditioned by water. The natural forces at work have inevitably shaped man's way of life in matters of clothing, shelter, diet, work, rest and sleep, customs and traditions.

Clothing

Man uses clothing in the desert, in winter and in summer, to protect his body from radiation, air temperature and dry wind during the day and from rapid cooling during the night. The customary clothing of desert dwellers consists of woollen robes (burnous) that cover the trunk, arms and legs, with a turban or cloth to cover the head and sandals to insulate the soles of the feet against ground heat. This simple garment is fitted loosely around the body, insulating it against warm and cold air temperatures whilst allowing air movement for evaporation. Tribal differences of clothing exist, but, on the whole, only the style and colours vary.

The Tuaregs of the Sahara are known to cover the entire head with cloth (litham), leaving only the eyes exposed. This practice of covering the nose and mouth, in addition to the head, may have developed through experience of trying to reduce pulmonary evaporation and to protect the lungs against dust, particularly in dry and stormy weather. In fact, the experiments of Adolph (1949) on men in the desert have shown that the thinnest, loosely worn white cloth gives adequate protection against warm air, sun and sand. Heat insulation and water conservation are the two main roles of clothing in the desert.

Shelter

No shelter of a temporary nature suits desert life better than the tent. Tents made of camel wool, goat's hair or leather serve as the habitations of nomadic and semi-nomadic tribes in the main hot deserts: mobility is their chief advantage. The raw materials for clothing and shelter come from the very animals that are the backbone of nomadic peoples' economy. Camels, sheep and goats serve as man's main companions. They are also his essential economic resources. His very sustenance and his movements in the desert depend upon them.

Nomadic life

The nomads are known for their simplicity, frugality, endurance, hospitality and love of freedom. They wander every season with their camels and herds in search of water and grazing. Their diet comprises milk and milk products, wheat, barley, dates, rice, sugar, tea, coffee, salt and occasionally meat (camel's or sheep's). They visit the towns in order to exchange what they have in animal products, rugs and cloth for the provisions that they need. According to Musil (1928), the Rwala bedouin eat twice a day. They take a light lunch before noon and their main meal, the supper, is taken shortly after sunset. On days when they undertake long marches they take grains of salt, a morsel of bread or milk as breakfast. They continue travelling until evening, without taking lunch. Milk and whole-wheat *ejs* (boiled crushed wheat) are their main nourishment. The Saharan Tuaregs are also known for their frugal food habits. Lhote (1937) said that a few dates and a bowl of sour milk suffice them to march for a whole day.

The favourite drink of nomadic people is tea or coffee. Water is drunk sparingly. Interesting social customs are associated with their habits of eating and drinking.

Tribal methods of social organization and communal existence prevail among nomadic populations, who generally lead a peaceful life. But tribal raids and wars still occur in certain regions. On the whole, nomadic life is influenced to a great extent by climatic changes and by the diurnal and seasonal natural cycles. Desert dwellers work or fight only when absolutely necessary, with ample relaxation and rest periods. Freedom, within tribal regulations, and austerity of life are part of man's behavioural adjustments to desert conditions.

There is a growing volume of literature on the life and customs of desert populations. The great deal of information that has been accumulated and facts related to biological sciences need to be examined.

PHYSIOLOGICAL STUDIES ON EFFECTS OF CLIMATIC ELEMENTS

The physiologist is interested in climate for reasons that Claude Bernard so well stated: 'L'être vivant fait partie du concert universel des choses, et la vie de l'animal n'est qu'un fragment de la vie totale de l'univers.' Climatic conditions are a vital part of man's daily existence.

With the advancement of science, empirical knowledge on biological phenomena led to experimental investigation. Lavoisier laid the foundations of energy metabolism. In his book entitled *Chaleur Animale et Bioénergétique*, Jules Lefèvre summarized, in 1911, the work of his predecessors and contemporaries. Energy metabolism and thermoregulation attracted the attention of scientists, but progress in the study of the heat-regulating mechanisms was slow.

Blagden (1775) reported that he exposed himself to dry heat at an environmental temperature of 121° C for 15 minutes without ill-effect. In 1905, Haldane observed men working in hot and wet mines and developed the concept of acclimatization as a physiological process. Most of the studies on the reactions of human organisms to environmental heat were conducted in artificially heated rooms. Field investigations on the subject are very sparse.

In recent years, many outstanding publications on man's physiological adjustments to environmental heat have appeared: Barbour (1921), Bazett (1927), Sundstroem (1927), Laurens (1928), Deighton (1933), Blum (1945) and Manchle and Hatch (1947) published reviews of literature in the *Physiological Reviews*; current developments from 1939 were reviewed in the *Annual Review of Physiology*; and significant books have been published by Kuno (1934), Benedict (1938), Dill (1938), Giaja (1938), Adolph *et al.* (1947), Dubois (1948), Winslow and Herrington (1949), Newburgh *et al.* (1949), Macpherson *et al.* (1960) and many others. While these publications advanced our knowledge of the adjustments that the human organism can make in a hot environment, field investigations on the adaptive mechanisms that indigenous people in hot climates developed naturally are totally lacking. The chronic effects of desert climate on man have not been investigated.

HOT DESERT CLIMATE

In the first part of this study we discussed at length the varying degrees of stress exerted on biota by the desert climatic elements of intense radiation, high air temperature, low atmospheric humidity, hot winds and extreme diurnal and seasonal fluctuations.

In the hot desert environment man's body is exposed to the following stresses:
1. Rapid gain of heat from the sun and the air;
2. Excessive sweating to lose surplus heat through evaporation;
3. Elimination of considerable amounts of salt in the sweat secreted;
4. Acceleration of the blood circulation to transport heat to the body surface for cooling;
5. Risk of undergoing 'voluntary dehydration' and salt depletion;
6. Danger of suffering from heat exhaustion and heat stroke;
7. Harmful effects of dust, glare and sunburn.

It is evident that the two main problems of man in a hot desert environment are (a) how to keep cool—that is, maintain heat balance—and (b) how to replenish the water and salt lost through sweat—that is, maintain water and salt balance. In the natural order of physiological adjustments, overheating of the body involves the mechanism of heat loss through evaporation, and its sequel of excessive sweating depletes the water and salt stores of the body.

It should be noted that whereas the mechanism of sweat secretion is automatic, requiring no conscious effort, the mechanisms of thirst and salt-craving serve only as dull signals that conscious effort is required to restore the body's needs for water and salt.

HEAT BALANCE

Body temperature

The normal rectal temperature of the human body is 37° C, with a daily cycle of a 1° C rise and fall. But there is a descending temperature gradient from the deep tissues to the skin. The temperature of the blood supplying the skin may vary by 20° C in the course of a single circulation, and a temperature difference of 10° C may be found between a resting muscle and an active one. According to DuBois (1948), normal persons may have temperatures varying between 36° C and 40° C, but rectal temperatures above 37·5° C should be investigated. The human organism can maintain a thermal balance of heat production and heat loss at different levels, within the limits of 4° C, without a pathological etiology. It is able to do this by exchanging heat and water with the environment.

The artificial means of protection which man uses enable him to bear a fluctuation of environmental temperatures of the order of 122° C; that is to say, he can live in the heart of the Sahara at an air temperature of 50° C, or in the heart of Siberia at an air temperature of −72° C. Without clothing, heating devices or shelter, however, he cannot tolerate for long environmental temperatures of 17° C or lower.

According to Herrington (1949), the tolerance limits of the human body to heat stress are:

Body temperature (37° C)	+6° C
on exposure to cold	−12° C
Increase in cardiac output	1·4–1·7 l./min
Pulse rate at rest	140/min
Pulse rate with exercise	180/min
Sweat secretion at rest	1800 g/h
Sweat secretion with exercise	3900 g/h

Adolph and associates (1947) have reported that the acceleration of pulse rate and the increase of cardiac stroke volume and rectal temperature may serve to measure the effects on man of heat, work, posture and dehydration under desert conditions. A rise in body temperature of about 2° F (1·11° C) is accompanied by an increase of 20 beats per minute in pulse rate. The rectal temperature of desert hikers rises about 0·55° F for every 1 per cent loss of body weight through depletion of body water.

Heat and water exchanges with the environment

To illustrate the extent of heat and water exchange with the environment, let us take as an example an adult man whose body weighs about 70 kg and who spends the day quietly in a comfortable environment.

It can be estimated roughly that such a man every hour metabolizes about 100 kcal of heat and 100 g of water. This means that he balances his daily income and expenditure of heat energy at 2400 kcal and his daily intake and output of water at 2400 g. We know that all his income of heat energy and water comes from the food and drink he takes from the environment. The routes of his heat elimination to the environment are, approximately:

	kcal	per cent
Conduction, convection and radiation	1600	67
Evaporation from the lungs and skin	600	25
Warming water and food ingested, air inspired and elimination of CO_2	170	7
Feces and urine	30	1
Total	2400	100

The routes of his water elimination are:

	grams	per cent
Urine	1400	58
Skin (cutaneous insensible perspiration)	620	26
Lungs (respiratory insensible perspiration)	290	12
Feces	90	4
Total	2400	100

The question now arises of what happens to such a man when he is placed in a hot desert environment.

Physiological adjustments to heat

Our present knowledge indicates that the naked human body under resting conditions adjusts to environmental heat as follows:

1. At the environmental temperature of 28° C, heat production is equal to heat loss, the skin temperature is close to 33° C and the internal temperature is kept constant at 37° C.

2. At lower ambient temperatures, of 22° C, vasoconstriction takes place, the skin temperature is lowered to 30° C and the metabolic heat production is increased. The body still manages to balance its heat production and heat loss at a higher level, and the internal temperature is kept constant at 37° C up to 2 cm from the skin surface. At environmental temperatures below 10° C, fatal cooling begins.

3. At higher ambient temperatures, of 30° C, vasodilatation takes place and the cutaneous circulation is increased. The skin temperature may rise to 34° C, but the body can still lose heat through sweating and greater evaporation, and the internal temperature is kept constant at 37° C.

At environmental temperatures above 30° C, the body relies increasingly on sweating for heat loss. At 40° C the skin temperature rises to 35° C, the body begins to store heat and the internal temperature may rise in proportion to the time of exposure. Heat gain through conduction, convection and radiation is added to metabolic heat, and the only way to lose heat is through evaporation. Every gram of water vaporized absorbs 0·58 kcal of heat, and to lose 100 kcal of heat 173 g of body water must be evaporated. The rise of blood temperature induces sweating centrally through the action of the hypothalamus. Sweating is also stimulated reflexly by local heat. The evaporative heat loss increases, and evaporative cooling takes place in proportion to the extent of wetted skin area, the skin-to-air vapour pressure gradient, and the rate of air movement.

Among the factors that influence evaporative water loss in the desert are air temperature, solar radiation, wind velocity, relative humidity, clothing, body size, muscular activity, emotional stress and degree of dehydration.

Heat gain and heat load

The body's reactions to ambient heat are vasodilatation, elevated skin temperature, increased blood volume and circulation, increased heart rate and systolic volume, increased respiration rate, elevated rectal temperature and heat production (at external temperatures of 30–35° C or over) and sweating.

Man protects himself against environmental heat as follows:

1. He avoids heat gain from direct solar radiation, indirect sky and ground radiation and hot air currents by using clothing, shade and shelter;

2. He reduces his heat production by avoiding muscular effort, by increasing relaxation and rest and by diminishing food intake (especially protein and fat);

3. He increases his heat loss through conduction, convection, radiation and evaporation by increasing cutaneous vasodilatation, by exposing a greater area of skin surface to cool atmosphere and air currents and by sweating.

Siple (1949) estimated that man's positive heat load in the desert may range from 0 to more than 300 kcal per square metre of body surface per hour. According to Blum (1945) the average solar heat load on a white man is 139 kcal/m²/h. This means that a man can gain three times as much heat as his resting metabolism produces. The amount of solar heat load depends on altitude, angle of the sun and dust and water-vapour contents of the air.

Adolph and his associates (1947) reported that at 100° F (37·8° C) in the desert the average clothed man in the sun gains 200 kcal/h from his surroundings (largely from direct and indirect radiation), or 2·5 times his metabolic heat production when at rest. At 110° F (43·3° C) his gain is more than 300 kcal/h, or 3·5 times his metabolic rate. If he removes his clothing, his heat gain increases by 100–140 kcal/h. If he works whilst exposed to 110° F (43·3° C) his working rate is reduced by 25 per cent compared with similar work performed in a cool environment at the rate of 300 kcal/m²/h.

WATER AND SALT BALANCE

Man's major physiological problem in the desert is how to balance his daily intake and expenditure of water and salt. The maintenance of equilibrium in the volume and composition of body fluids is essential for normal function. Heat balance is inconceivable without the acquisition and conservation of adequate amounts of water and salt.

The daily intake of water and salt is conditioned by the following factors: (a) availability of adequate provisions in the environment, (b) individual habit and social custom and (c) the physiological mechanisms of thirst and appetite. The kidneys and the skin take care of the elimination as well as conservation of water and salt, but they can do nothing for their replacement. The main concern of man in the desert, therefore, is to avoid dehydration and salt depletion.

Body fluids

It is estimated that the body of an adult man weighing 70 kg contains 46 l. of water, or 66 per cent. This water exists in three interchanging compartments; namely, 3 l. as blood plasma, 14 l. as interstitial fluid and 29 l. as intracellular fluid.

As the membranes that separate these compartments permit the free passage of water, the osmotic pressure of body fluids is kept at a constant level. The total concentration of electrolytes is the same in different parts of the body but their composition varies because of the differential permeability of the membranes to solutes. A 0·9 per cent solution of sodium chloride is considered to be isotonic, and its osmotically effective concentration is 0·310 molal, similar to that of serum, 0·309 molal, and that of cells, 0·307 molal. Water of blood plasma and of interstitial fluid is referred to as extracellular water, and intracellular water is that which exists within the cells.

Water and salt excretion

Sodium and chloride ions predominate in extracellular fluid, but they are lacking in intracellular fluid. Thus, when sodium chloride is depleted or water is added to extracellular fluid there is a shift of water from the extracellular to the intracellular fluid in order to maintain osmotic balance. Conversely, when water is depleted or sodium chloride is added to extracellular fluid there is a shift of water from intracellular fluid to extracellular fluid for osmotic equilibrium.

It is the role of the kidneys to regulate the volume and composition of the body fluids by altering the volume and composition of the blood which flows through them (a quarter of the cardiac output). They perform this task by excreting the excess salt and water in the body after its needs are satisfied. The function of the kidneys is regulated by circulatory, hormonal and possibly neurogenic factors.

According to Adolph (1949), the body contains about 165 g of sodium chloride, and 10–15 per cent of this amount is metabolized daily. To eliminate excess salt in urine, about 50 ml of water are needed for each gram of sodium chloride excreted. When 2 g of sodium chloride per litre of urine are excreted, it is an indication that the body has an adequate supply of salt: a daily output of 2 g of sodium chloride in the urine is considered to be normal. The waste products to be excreted amount to 35 g daily, and to excrete this amount in maximum concentration the kidneys require about 500 ml of water.

Fluid movements under thermal stress

According to Bass and Henschel (1956), body fluids respond to thermal stress as follows:

1. Exposure to excess heat results in: (a) hemodilution, with contraction of the interstitial fluid volume and expansion of the plasma volume; (b) decrease of glomerular filtration and sodium chloride and water excretions; and (c) urinary response for the re-establishment of a more normal ratio of interstitial volume to plasma volume.

2. Exposure to excess cold results in: (a) hemoconcentration, with contraction of plasma volume and expansion of interstitial volume; (b) increase in urinary water and salt excretions; and (c) urinary response for the re-establishment of a more normal ratio of interstitial volume to plasma volume.

Smith (1956) supports the view that the water balance of the body is so regulated as to maintain the osmotic pressure of the plasma in a steady state. Its coefficient of variation is only 1·2 per cent, the smallest of the physiological variables. The regulation of osmotic pressure has priority over that of volume.

Mechanisms of water and salt conservation

According to Strauss (1957), phylogenetically speaking, the mechanism of salt conservation developed first, as life migrated from the sea to fresh waters, and the mechanism of water conservation developed later, as life ascended to dry land. Renal control of salt and water excretion, thirst and appetite for salt are regulating mechanisms for maintaining water and salt balance in the body. The neurohypophyseal secretion of antidiuretic hormone (ADH) enables man to excrete urine four times as concentrated (hypertonic) as his body fluids, while certain desert mammals manage to excrete urine up to 17 times as concentrated.

ADH is released into the blood by the hypothalamico-hypophyseal system in response to the diminished volume of extracellular fluid (particularly plasma) and to unpleasant and painful stimuli. The release of ADH is inhibited—and water diuresis results—when osmotic pressure is decreased, with the expansion of extracellular and intracellular fluid volumes. Alcohol inhibits the release of ADH, whereas acetylcholine, nicotine and sedative, narcotic and anesthetic agents provoke its release. Renal excretion of water is controlled by the supraoptico-hypophyseal system.

Renal salt excretion responds to body-fluid volume changes by increased or decreased tubular reabsorption of sodium. Sodium excretion is increased when the total body water or extracellular fluid volume is increased, and vice versa. Aldosterone is known to have a salt-retaining factor. The existence of a 'salt-excreting' hormone is suspected. The hypothalamus may have an area which regulates renal salt secretion.

SWEATING

Sweating is a mechanism of cooling, other than insensible perspiration, when the body is overheated, and evaporation is the only way to lose heat at environmental temperatures above that of the skin. Every gram of sweat evaporated on the skin surface removes 0·585 kcal from the body. Hence one litre of sweat may remove about 600 kcal, which is a considerable amount of heat to enable man to tolerate higher temperatures.

Sweat glands

According to Kuno (1956), man's capacity for sweating is highly developed in comparison with other animals. Some animals, having no sweat glands, use panting, salivation and wetting of body parts for cooling. It is estimated that man's body has about 2·28 million sweat glands. These glands may secrete more than 3 kg of sweat per hour, for 3–4 hours, in men acclimatized to heat. There are two types of sweat gland: (a) the eccrine glands that respond to thermal stimulation and are distributed over the entire surface of the body and the eccrine glands that respond to mental stimulation and are distributed on the palms and soles, and (b) the apocrine glands that respond to thermal as well as mental stimulation and are distributed on the axilla.

Recent studies indicate that the thermal sweating centre is located in the hypothalamic region and the mental sweating centre probably in the cerebral cortex. The secretory nerves of sweat glands are mainly sympathetic and partly parasympathetic. Sweating is stimulated by pilocarpine and acetylcholine and abolished by atropine.

Sodium chloride

The total solid constituents of sweat comprise 0·3–0·8 per cent, with chloride and sodium ions predominating in concentrations a little lower than those of plasma (in mg per cent, sweat contains Cl^-: 320 and Na^+: 200; plasma contains Cl^-: 360 and Na^+: 340). Sweat is also known to contain other constituents of plasma and vitamins in insignificant amounts.

The concentration of sodium chloride in the sweat varies from very low levels to the isotonic level, the average being 0·3 per cent. The following factors influence its concentration: (a) individual variation, (b) the rate and duration of sweating, (c) the level of salt intake, (d) the skin and rectal temperatures, (e) the state of acclimatization and (f) the conditions under which sweating takes place.

When salt intake is low, the kidney cuts down its salt concentration and output, and this is followed by a reduction of the salt content of sweat. The sweat glands are able to do osmotic work to secrete hypotonic sweat.

According to Kuno (1956), the chloride concentration of sweat rises in proportion to the rate of secretion, and it may even approach the blood plasma level. In prolonged sweating, without water intake, chloride continues to appear in sweat in small concentrations. Naturally, every litre of sweat secreted leaves behind excess salt in the extracellular fluid. It is observed, however, that the chloride concentration of the serum remains unchanged up to two litres of sweat loss. Only when 4–5 l. of sweat are lost will a rise in serum chloride concentration take place.

The mechanisms which protect the level of chloride concentration in the serum and the extracellular fluid are as follows: (a) the kidneys excrete urine containing a high chloride concentration; (b) chlorides accumulate in the skin around the actively secreting sweat glands, returning gradually to the bloodstream along the lymph channels; (c) antidiuretic hormone of the pituitary body is discharged as dehydration takes place; and (d) water shifts from the intracellular fluid to the extracellular fluid to maintain osmotic equilibrium.

When sweating is excessive, and no water is taken, more than 30 g of salt may be lost in a day. Although the normal daily intake of sodium chloride is 3–10 g, a healthy person may tolerate 35–40 g. In profuse sweating, in order to replenish the body's stores it is necessary to take adequate amounts of water and salt (0·2 per cent salt solution may be ingested).

Shotton (1954) stated that human beings living in a desert climate may regularly drink underground water with a salinity of 3000 parts by weight of sodium chloride per million of water. This is one way of satisfying their salt needs.

Sweating rate

Winslow (1949) found that in the nude man, at rest in a hot environment, sweating is initiated when the skin reaches the critical temperature of 34·5° C. The rate of sweating is influenced by the level of skin and body temperatures.

Adolph (1949) reported that in the desert an increase of 1° F (0·56° C) in air temperature increases the sweating rate by about 20 g per hour, while an exposure to direct sunlight increases the sweating rate by 200 g per hour, or an air temperature equivalent of 10° F (5·55° C). Man working at the rate of 300 kcal/m^2/h in a cool environment, when exposed to an air temperature of 110° F (43·33° C), has his working efficiency reduced by 25 per cent, which is equivalent to a water deficit of 2·5 per cent (1·8 l.). If the same man is initially exposed to 110° F (43·33° C) and dehydrated by 2·5 per cent of his body weight, his working efficiency is reduced by 50 per cent. Most of the sweat in the desert evaporates from the body surface imperceptibly (invisible sweating), and man's evaporative rate rarely reaches 1500 g per hour.

The following data indicate the sweating rates of men in the desert at an air temperature of 37·8° C:

> 300 g/h when sitting in the shade and clothed;
> 900 g/h when walking in sun, clothed;
> 1150 g/h when walking in sun, nude.

Men engaged in varied military activities in the desert showed an average sweating rate of 4·1 l./day. Night work saved 3·5 l. of water per day. Maximal total losses of sweat of up to 10–12 l./day were exhibited by men working in the desert.

Weiner (1954) estimated that under desert conditions the human body, kept wet at a skin temperature of about 36·5° C and air speed of 8 km/h, could lose heat at the rate of about 500–600 kcal/h. His convective heat load might be 100 kcal/h, his radiation heat load 150 kcal/h, and there remain 250 kcal/h for metabolism and work. Hence the maximum capacity of the human body to maintain heat balance demands an output of about 1 l./h of sweat in order to lose heat at about 500 kcal/h. Tests on Europeans have shown that this rate of sweating (1 l./h) can be maintained by the sweat glands for 4–6 hours.

Ladell (1953) reported that the water lost in sweat comes from the extracellular fluid and that a negative balance of 2·7 l. can be incurred before water is drawn from the intracellular fluid. He observed in Iraq that a subject remained in good condition after sweating 440 ml/h and abstaining from water for eight hours. Symptoms of circulatory failure did not appear until after 12 hours. Also, Ladell, Waterlow and Hudson (1944) reported that moderately active men sweated 0·5 l./h in the shade in the Arabian desert.

According to Kuno (1956), sweating observed in overheated factories and among active soldiers in the tropics is 10–15 kg/day. The rate of sweating is lower under dry heat than under wet heat. As the sweat evaporates readily under dry heat, the skin temperature rarely rises above 35° C, but under wet heat it may rise much higher. Excessive exposures to heat may cause 'sweat-gland fatigue', when the sweating rate declines rapidly.

ACCLIMATIZATION

Acclimatization is the process of conditioning the body's latent capacities for improved adjustment to climatic stresses. According to Adolph (1949), acclimatization to desert conditions can be acquired within a week through training, and its effects endure for several weeks.

Men acclimatized to desert heat show an improved heat regulation. Their pulse rate, pulse pressure, skin and body temperatures, salt concentration in sweat and basal oxygen consumption are decreased. Their rate of evaporation and heat tolerance are increased. They have an increased capacity for muscular effort.

In men exposed to desert heat, individual differences are found in circulatory performance, sweat formation and salt conservation. These differences can be detected before and after acclimatization, indicating that some factors of adaptation may be inherited.

Kuno (1956) reported that the natives of the tropics have a larger number of active sweat glands, and they are therefore able to sweat more intensely. For example, the Russians have 1,886,000 sweat glands, while the Filipinos have 2,800,000 sweat glands (ratio 2 : 3). Sweat glands may be activated within two years after birth, not later. An individual from a race in the temperate zone, therefore, if born in the tropics, can acquire a sweat gland equipment as fully developed as that of the natives. It has also been observed that the maximum chlorine concentration in the sweat of natives of the tropics is one-third that of the Japanese. Men living in a hot environment show, for a standard work performed, reduced chloride concentrations in their sweat and urine. They are able to balance their salt intake and output on a lower level.

Bass and associates (1955) reported that acclimatization to heat is accompanied by an isotonic expansion of the extracellular fluid volume. According to Peters (1950), the body responds to loss of water with diminished excretion of sodium and chloride ions in the urine, despite their rising concentration in the serum. This is called the 'dehydration reaction' and is accompanied by a cell-water shift to the extracellular compartment.

Kuno (1956) and Robinson (1949) support the view that acclimatization may be attributable to the stimulation of the activity of the adrenal cortex. The production of adrenocortical hormones may be increased in repeated exposures to high temperatures, enabling the sweat glands to secrete sweat with a decreased chloride concentration. Actually, the administration of desoxycorticosterone acetate (DCA) leads to a lowering of salt concentration in the sweat and urine of acclimatized as well as unacclimatized men. With the progress of acclimatization, sweat becomes more dilute in chlorides and the ability to sweat increases. In acclimatized working men, sweat may be secreted at the rate of 4 l./h for short periods.

DESERT HAZARDS

Among the supplies essential for life and work in a desert environment are water and salt. Adolph and his associates (1947), who determined the survival limits in the desert, stated that the human body cannot be trained to economize water. There is no 'acclimatization' to lack of water. Man's water requirements can be reduced by the proper use of shade and clothing and by avoiding any activity that increases heat gain or prevents heat loss, but any water deficit in the body must be restored. This does not mean that there are no natural individual differences of endurance to dehydration. The diary of a group of men lost in the Nubian Desert in 1959 mentioned that their native guide was 'a patient man, who spoke little, ate little and drank only once a day'.

Dehydration

In the hot desert a man may lose about 1 l. of sweat per hour. If he continues to sweat, without drinking water, until he loses about 5 per cent of his body weight, he will suffer serious ill-effects. If he becomes dehydrated to the extent of losing 10–12 per cent of his body weight, his life is endangered because undue hemoconcentration leads to circulatory failure and the quick rise of body temperature may result in death. The survival limit is 20 per cent of body weight. The dehydration risk of children is greater than that of adults.

In the fight against heat, water balance is disturbed for the sake of maintaining heat balance. The physiological disorders experienced in the desert are directly or indirectly related to the factors of heat, water and salt. Drinking frequently is advisable. Night work and travel, clothing and shelter help to reduce water expenditure. Adolph (1949) estimated that for every 20-mile walk one gallon of water is required by night and two gallons in daytime. The average distance that a man can walk without water in summer is about 20–25 miles. Body weight and plasma concentration are reliable criteria of water balance in active men.

Thirst mechanism

As early as 1856, Claude Bernard observed that thirst is not a local sensation but a generalized feeling associated with the state of dehydration. The studies of Adolph (1949) showed that thirst sensations are not enough to induce a man to drink all the water he needs. In the desert, when sweating is rapid, man may drink only half the amount of the water he loses, allowing himself to live in a state of 'voluntary dehydration'. He takes most of the fluid he needs at rest periods with meals. He may take as long as 24 hours to replenish his water stores. Therefore, the thirst mechanism is not an adequate safeguard against water deficit. This tendency to voluntary dehydration, owing to the inhibition of thirst sensations, of up to 2–5 per cent of body weight, may hamper working capacity. Voluntary dehydration may cause increased pulse rate and rectal temperature.

Dehydration exhaustion

According to Adolph (1949), when the water deficit reaches 5–6 per cent of the body weight, 'dehydration exhaustion' sets in, the easily detectable symptoms of which are high pulse rate and high rectal temperature. Drinking water relieves dehydration distress within a few minutes, but in an advanced state of dehydration the following symptoms appear: pulse rate and rectal temperatures increase; breathing rate accelerates; blood becomes concentrated and its volume diminishes; gastrointestinal upsets, nausea, lack of appetite, difficulty in muscular movements, numbness and emotional instability occur. These symptoms are aggravated when dehydration reaches 6–10 per cent of body weight. A water deficit of 11–20 per cent of body weight causes delirium, deafness, dimness of vision, swollen tongue, anuria and perhaps death. A dehydration of 6 per cent of body weight is equivalent to a heat exposure which raises the rectal temperature by 1·2° F (0·67° C) and which increases the pulse rate by 40 per cent.

Other risks to which men in the desert may be exposed are:

1. *Heat stroke:* The symptoms are high body temperature, delirium, diminished sweating, etc. The remedy indicated is quick cooling of the body, fluid ingestion and rest.

2. *Heat exhaustion:* The symptoms are faintness, rapid and weak pulse. The remedy indicated is fluid ingestion and rest.

3. *Heat cramp:* The symptoms are pain and muscle spasm. According to Le Breton (1959), heat-cramp patients suffer from dehydration, low concentration of sodium and chloride ions in the plasma, little or no chlorides in the urine and increased serum protein concentration. The injection of physiological serum and the administration of salt provide relief and recovery.

CONCLUSIONS

The scientific investigation of the effects of desert climate on man has been started only recently and our present knowledge on the subject is limited to man's physiological responses to certain climatic elements. The chronic effects of desert climate on indigenous populations remain to be investigated.

Man is not an exception to the laws of nature. If plants and animals have developed adaptive mechanisms, surely man has done likewise. Future research will reveal the nature of his adaptations.

What the nomads accomplished with their elementary tools can surely be improved by modern science and technology. The fertilization of desert areas is already on the programme of all the governments of the world community of nations. The days of the spear and boomerang have passed. The deserts with all their resources, physical, animal and human, must receive the best attention of scientists and authorities everywhere.

Already the efforts of many nations are being organized, and the praiseworthy contribution of the UNESCO to the study and development of deserts is known. The proceedings of the International Symposium on Desert Research (1953) and the findings of the Symposium on the Biology of the Deserts (1954) describe the research carried out in different countries in order to find solutions to desert problems.

The thesis of Monod (1953) concerning the studies that need to be undertaken to develop a 'desert biology' constitutes a challenge to men of science throughout the world.

12
SUMMARY AND CONCLUSIONS

ADAPTATION

After a brief introduction to the subject of animal adaptation to desert we described the desert environment and reviewed the varying degrees of stress which its rigorous conditions exert on biota. In the first chapter sufficient evidence was cited to show that, in their struggle for survival, desert plants and animals have, in the course of evolution, developed mechanisms of adaptation which can be characterized as ethological, ecological, physiological, morphological and genetic.

COMPARATIVE STUDY OF RODENTS

In order to discover the adaptive and regulative mechanisms of thermo-regulation that two different species from two different environments developed, we have undertaken a study of the jerboa (*Dipus aegyptius* or *Jaculus orientalis*), a rodent of the Egyptian Western Desert, and the white rat (Wistar), a rodent of temperate regions.

THE JERBOA AND ITS HABITAT

In Chapters 2 and 3 results were given of our investigation on the jerboa's biological history, distribution, morphological features, desert habitat, burrow, ecoclimate and mode of life in the desert. We have found that nocturnality and underground living are the jerboa's partial solutions to the problems of insufficiency of water and food and intensity of heat and radiation in the desert.

THE JERBOA IN THE LABORATORY

As we have introduced the jerboa to laboratory life for the first time, all necessary measures were taken to condition the animal properly and study its habits of feeding, burrow-building, rest and sleep and behaviour. We have, therefore, reached the conclusion in Chapter 4 that, if the essential conditions for its reproduction could be realized, the jerboa could become an excellent laboratory animal and new studies could be undertaken on this desert species.

THE JERBOA COMPARED WITH THE RAT

In Chapter 5 we described the results of our observations on the comparative growth of the jerboa and the rat and the effects of wet and dry diets on their body weights. We have found that the body development of the jerboa takes twice as long as that of the rat and its longevity is double that of the rat.

SUMMARY AND CONCLUSIONS

Furthermore, the jerboa is able to live on dry diet for 1–3 years, whereas the rat cannot support dry diet for more than three days.

EFFECTS OF DIET

The effects of diet on excretion, body-water content and spontaneous natural activity have also been examined. Our results, given in Chapter 6, indicate that on wet diet the jerboa eats less than the rat per 100 g of body weight, its excrements of feces and urine are quantitatively almost half of those of the rat, it has 4·2 per cent more water and 20·9 per cent more fat in its body than the rat, and its spontaneous activity during the daily 24 hours is less than that of the rat.

Put on dry diet, the jerboa eats less and excretes less, its urine becomes more concentrated and acidic and it can continue to live thus for 1–3 years without any organic lesion; whereas the rat does not tolerate dry diet, as it ceases to eat after the third day and lesions in the intestinal and urinary tracts appear towards the sixth day. There must be a profound difference in the renal function of these two animals, and this matter needs further investigation.

A jerboa on continuous dry diet has 13·6 per cent less fat and 4 per cent less water in its body than one on wet diet, and its 24-hour natural activity is feebler, thus economizing energy and water. It is interesting to note that even the jerboa on continuous dry diet has 7·3 per cent more fat and 0·2 per cent more water in its body than a rat on wet diet.

BODY TEMPERATURE

Chapter 7 gives our findings on the body-temperature variations of jerboas and rats in relation to external temperatures. We have observed that the jerboa is better adapted than the rat to high temperatures; it resists ambient temperatures for which the rat is under stress, and it is thus that at 35° C the rat's body temperature reaches 38·7° C and the jerboa's only 37·7° C. The differences are accentuated when the external temperature reaches 40° C (39·4° C for the jerboa; and 41·1° C for the rat). Dry diet diminishes the thermoregulatory power of the jerboa; it shows signs of stress at a temperature lower than that observed for a jerboa on wet diet. The jerboa enters a state of deep sleep and salivates profusely at high temperatures, whereas the rat is agitated.

ENERGY METABOLISM

We have also investigated the energy metabolism in relation to external temperatures from 0° C to 45° C of jerboas on wet diet, rats on wet diet and jerboas on dry diet. The results of this series of experiments appear in Chapter 8. They show that the temperature of thermal neutrality in these three cases is close to 30° C. The basal metabolism of the jerboa is netly inferior to that of the rat, in spite of its small size (3·98 kcal/kg/h for the jerboa and 6·15 kcal/kg/h for the rat, both on wet diet).

We noted that when the external temperature falls the metabolism of the

SUMMARY AND CONCLUSIONS

jerboa rises to the level of that of the rat. The oxidative processes due to thermoregulation are higher—14 instead of 13 kcal/kg/h—because here it is the dissipating surface that regulates the level of oxidation under the influence of reflexly secreted hormones.

The respiratory exchanges increase faster in the rat than in the jerboa when the ambient temperature exceeds 30° C. The factors responsible for the jerboa's resistance to heat have been analysed:

1. The basal level of oxidation is lower;
2. Total absence of muscular movement—state of lethargy close to 35° C;
3. Body position favourable;
4. Abundant salivation accompanied by the evaporation of saliva secreted and spread on the body surface;
5. Resistance to heat is equally favoured by the adjustment of the mechanisms of thermolysis, which are put in operation to save every loss of water not entailing an efficacious vaporization; namely, the dissipation of calories liberated by the oxidation.

INSENSIBLE PERSPIRATION

The results of our researches on the action of thermolysis in relation to external temperature, in jerboas and in rats, are presented in Chapter 9. We found that at thermal neutrality insensible perspiration (water vaporized at the level of the skin and lungs) is weaker in the jerboa on dry diet than in the jerboa on wet diet. It is, on the other hand, about twice as high in the rat as it is in the jerboa (1·136 g/kg/h against 0·66 g/kg/h); that is to say, the percentage of thermolysis resulting from this insensible perspiration is netly higher in the rat. When the temperature drops from 30° C to 0° C, this percentage passes from 9·1 to 11·4 in the rat, from 4·6 to 10·2 in the jerboa on wet diet, and from 3·7 to 8·7 in the jerboa on dry diet. If the temperature rises above 30° C, the inverse phenomenon is observed and the jerboa on dry diet economizes to the maximum its reserves of water.

We have also calculated the ratio of H_2O/O_2 from the measurements of water vaporized and oxygen consumed, and arrived at the classical facts in the case of the rat—the ratio increases regularly from 0° C to 30° C and then shows a considerable rise between 30° C and 40° C, since at 40° C all the heat produced is dissipated by vaporization of water. We have explained above that the reasons for the jerboa's resistance to high temperatures, in spite of the relative privation from water to which it is subjected, are the mechanisms of adaptation that were installed in this species during the course of its evolution in the desert climate, mechanisms among which the reflex sleep (estivation) provoked at high temperatures is the most interesting.

THERMOREGULATION

In Chapter 10 we recapitulated the experimental data of our study on the thermoregulation of the jerboa and rat and presented graphically their heat balance and water balance at the ambient temperature of 25° C. These

SUMMARY AND CONCLUSIONS

balance sheets indicate clearly the economy in energy and water which the jerboa on dry diet is capable of making because of its mechanisms of adaptation to the trying factors of the desert environment; namely, the insufficiency of water and food and the intensity of heat. The rat, being an animal of temperate regions, has a higher energy metabolism, does not support dry diet and does not tolerate high ambient temperatures.

The data we obtained on the thermoregulation of the jerboa, a desert rodent, need to be completed by researches on another species, *Jaculus jaculus*, of smaller size and adapted to hotter and drier desert than the one of the Egyptian littoral. The problems to investigate are: composition of body fluids; role of the glands of internal secretion; reproduction in the laboratory; mechanisms of renal secretion when the animal is on dry diet; lethargic sleep, 'estivation' and salivation, the reflexes which are produced when external temperatures rise above 35° C.

MAN'S ADAPTATION TO DESERT

In Chapter 11 we reviewed the literature on man's adaptation to desert climate from two aspects:
1. Field observations on climate and human race,
2. Physiological studies on the effects of desert climate.

It must be noted that most of the recent studies are limited to man's physiological responses to climatic elements and that a study of the chronic effects of desert climate on desert dwellers is totally lacking. We have also discussed the present findings on man's heat and water exchanges with the environment, on water and salt balance, on sweating and acclimatization and on desert hazards. It is evident that the growing interest in desert problems will stimulate desert studies that will lead to the development of a 'desert biology'.

GENERAL CONCLUSIONS

Our studies on the jerboa, rat and man, and the data presented in the foregoing chapters on desert ecology and the effects of desert climate on biota, indicate the following:

1. Both plants and animals have developed, in the course of their evolution, adaptive mechanisms to withstand the rigorous conditions of the desert environment; these may be classified as behavioural, ecological, physiological, structural and genetic. The jerboa is a good example of a desert species which developed all these mechanisms.

2. Scarcity of water, insufficiency of food and intensity of radiation and heat are the primary factors of desert stress on biota.

3. In the nature of all living organisms there exist latent capacities for adaptation, which only rigorous environmental conditions can help to develop.

4. Animals have more ways of adaptation to climatic stress than plants because of their mobility, and men have better possibilities of overcoming adverse climatic conditions than animals because of their technological advancement.

5. Man's adaptation to desert climate is physiologically possible and technologically feasible.

6. The development and exploitation of deserts require scientific investigation and technological mobilization.

7. The ecology, biology and physiology of desert life can contribute to the conquest of those immense wasted regions of the earth, the deserts, and to the development of their human and natural resources.

Realizing that the scientific investigation of desert problems is yet at its beginning, we trust that this modest study will stimulate further research on life in the deserts. We shall pursue our own studies, inspired by the high example of men of science who are dedicated to the arduous task of discovering the secrets of nature.

APPENDIXES

Appendix 1

TECHNIQUE

The fundamental purpose of this study is to establish the characteristics of the body temperature and energy metabolism of the jerboa, a desert rodent, in comparison to the white rat, a non-desert rodent.

In order to obtain comparable data, the animals were subjected to identical experimental conditions, and one factor at a time was varied. Adult animals, aged about one year, were used for all experiments. Their body weights varied little.

In establishing the plan and technique of the experiments, we followed the general principles of research proposed by Fisher (1951), Wilson (1952) and Federer (1955).

Our experimental material permitted the simultaneous study of three to six animals at a time. A detailed description of the material and technique used is given in this appendix.

LABORATORY FOR RESEARCH

Our laboratory comprised three rooms: a room in which the animals lived, a room for nutritional studies and a laboratory for experiments.

Animal room

The jerboas and white rats lived here in separate cages. We followed the recommendations of Worden (1957) with regard to the care to be given to animals and their conditions of existence.

The room is equipped with cages, a heating system, thermohygrometers, an actograph and the other materials used in our study.

Each cage consists of a galvanized-iron box 35 cm long, 24 cm wide and 24 cm high. It is closed by a lid of 35 cm by 24 cm, made of wire netting with a 7 × 7 mm mesh to permit adequate ventilation. Inside and at the back of the cage a section 9 cm high and 13 cm wide serves as a refuge or nest. On the top of this section food and water receptacles are placed. The floors of the cages are covered with a thick layer of clean sand, which is renewed every two or three days to maintain cleanliness.

Each cage contains one or two animals. The cages are arranged in rows, on two levels, at a height of one metre from the floor, in order to facilitate access. This animal room can easily accommodate 30 jerboas and 30 rats. A large communal cage periodically provides the animals with a change and more freedom. *Figure 26* shows the jerboas' and white rats' cages.

Nutrition room

Here the nutrition and excretion experiments are conducted. The room contains a metabolism cage, animal weighing box, drying oven and other equipment.

Metabolism cage: This cage is 2·10 m long, 0·40 m wide and 0·40 m deep, as illustrated in *Figure 27*. It is composed of six compartments, in order to

APPENDIX I

enable the simultaneous study of six animals. The top, front and sectional partitions are grilled by wire netting. Each compartment has two doors, one at the front and the other at the top. The floor of each compartment consists of two separate movable frames of galvanized-wire netting placed on

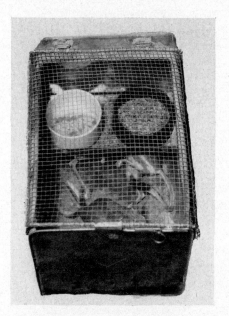

Figure 26. The cages for the white rats and jerboas.

top of each other. The top frame, whose openings are 7 × 7 mm, allows the urine and feces to pass, while the lower frame, whose openings are 3 × 3 mm, allows only the urine to pass. Underneath these two frames is placed a large funnel, 30 cm in diameter, which drains the urine directly into the collecting bottle. The animals can be placed singly or in pairs in each compartment. This cage enables accurate study of the animal's food consumption and excretion.

The animal weighing box: Made of plastic materials, the box is composed of two pieces: (a) the bell-shaped top cover, 17 cm in diameter and 8 cm in height; (b) the tray-shaped base, 20 cm in diameter. The cover contains openings for ventilation. As the weight of the box is known, it is easy to isolate the animal in the box and weigh it without causing any excitation.

The drying oven: A double-walled chamber, 25 cm each side. The space between its walls is filled with water and glycerine in order to maintain the temperature at 105° C.

Figure 27. Metabolism cage.

Experimental room

This room is equipped with an air-conditioning chest, confined air containers, water pump, cylindrical cages, Carpenter's apparatus for gas analysis, stand for air-sampling bottles, precision scales, precision barometer, dial thermohygrometers, the dry and wet bulb 'whirling hygrometer', thermometers and other equipment.

The air-conditioning chest: A chest 1·42 m long, 0·75 m wide and 1·05 m high. It has double walls separated by a space of 10 cm. The empty space between the two wooden walls is filled with powdered cork. In the interior of the chest are installed a thermostat and a ventilator to control the temperature. The cooling unit is placed outside the chest, but the electrically controlled heating unit is placed inside it. The chest has sufficient space to hold three containers, placed at equal distances from each other to allow free circulation of air around them. The chest temperature can be regulated between 0° C and 50° C and maintained constant for several hours with a variation of only 0·5° C.

Confined air containers: Made of galvanized iron, each container is about 30 cm in diameter and 50 cm in height, with a neck 10 cm in diameter

APPENDIX I

and 26 cm in height. In the interior of each container are installed an electric motor and an electric lamp. When the three containers are placed inside the chest, the extremities of their necks pass through the holes in the cover and extend outside the chest. Each container has its own thick rubber stopper. Through the centre of the stopper two glass tubes are inserted: (a) the tube for taking a sample of confined air; (b) a mercury manometer to indicate the pressure of the confined air. The stopper ensures hermetic closure of the container. The volume of the container is calculated from the weight of water it can contain (taking into account the temperature of the water). The capacities of the three containers, thus calculated, are 35·853 l., 35·705 l. and 35·614 l.

Cylindrical cages: Three cylindrical cages, corresponding to the three containers, are placed inside the containers before the experiment begins, and the air in the containers is brought to the desired temperature. When the experiment is started, each animal is made to slide through a carton tube into the container's cage and the stopper is fixed.

The cylindrical cage is made of galvanized iron. It is 8·5 cm in diameter and 60 cm in length and consists of two sections, joined together in the middle. The upper half is made of galvanized-wire netting, the holes of which are 7 × 7 mm. The lower half is made of galvanized-iron sheet, and contains a glass funnel and a bottle. The animal rests in the upper section, thus occupying the centre of the container. Its excrements, if any, are collected in the lower section. The cylindrical cage is isolated from the walls of the container by cork plates.

After two or three trials, the animal becomes accustomed to the cylindrical cage and it remains in it quietly throughout the experimental period. At 15 cm from the place where the animal rests, a metallic stopper blocks the exit, preventing attempts to escape (*Figure 28*).

When the experiment is terminated, the cylindrical cages are removed from the containers and the animals are returned to their normal cages.

The air in the containers is renewed as follows: First the ventilator in the container is started, and then the container's air is absorbed by means of a water pump attached to a metallic tube that passes through the stopper and extends as far as the bottom of the container. The pure air enters through a small hole, also through the stopper. By opening and closing this hole, the circulation of fresh air can be accelerated owing to the partial vacuum created. This procedure of air evacuation is repeated before and after each experiment. The container's air is analysed to guarantee its purity.

Air-sampling bottles: A stand carries three air-sampling bottles and a common reservoir of mercury. Each bottle has a capacity of about 150 cm^3 and is provided with a two-way tap at the upper end and a one-way tap at the lower end. Rubber tubes connect the lower-end taps to the reservoir of mercury.

After the experiment the samples of confined air are withdrawn from the containers into their corresponding bottles. The analysis is made the following day (see *Figure 28*).

Carpenter's air analysis apparatus: This apparatus is well known and needs no description. To control its proper functioning, pure air is periodically

TECHNIQUE

analysed. The analyses are made according to the method recommended by Haldane and Graham (1935).

The various apparatuses described above—the air-conditioning chest, confined-air containers, cylindrical cages, air-sampling bottles and Carpenter's apparatus—enable precise measurement of indirect calorimetry by the method of air confinement. The duration of the experiments is controlled in such a way that the carbon dioxide in the confined air does not exceed one per cent.

Figure 28. Cylindrical cage and sampling bottles.

ANIMALS STUDIED

The animals used in all the experiments in this study were:

(a) *Dipus aegyptius* or *Jaculus orientalis*, referred to as the jerboa.

(b) The white rat (pure race from the Institute of Wistar), referred to as the rat.

Jerboas were brought from the desert and habituated to laboratory life for several months before being subjected to experimentation. Rats (Wistar) were born and raised in the laboratory. All jerboas and rats studied by us lived in the animal room under the same conditions.

APPENDIX I

NUTRITION AND EXCRETION

The effects of a dry diet composed of dry wheat grains and of a wet diet composed of dry wheat grains plus lettuce, on the food consumption and the feces and urine excretion of jerboas and rats, were studied for periods of six days at a time, making use of the metabolism cage described above.

The feces and the urine excreted by the animal each day were, according to the type of experiment conducted, either weighed without drying or weighed after drying in the oven at 105° C until they reached a constant weight.

Similarly, the water content of food, feces and urine was determined by the difference between the fresh weight and the weight of dry residue after desiccation at 105° C.

Jerboas on a dry diet of dry grains of wheat and barley lived in separate cages for periods of more than one year, and their weights were recorded regularly.

The methods used in the other experiments were described in detail in the appropriate chapters.

BODY TEMPERATURE

Apparatus

For the study of the body temperature of jerboas and rats the following apparatus was utilized:

(a) An anesthesia bell (a transparent glass vase 20 cm in diameter and 11 cm in height) is supplied with an elastic rubber band encircling the rim of its opening. To make the bell airtight it is placed on a glass plate 30 × 30 cm.

(b) A clinical thermometer with a sensitivity of half a minute and a mercury bulb 5 mm long and 3 mm thick, to permit easy penetration into the rectum of the animal, were used.

Method

The method adopted for taking the rectal temperature was as follows:

(a) The anesthesia bell, including cups containing the ether, is prepared.

(b) The animal is removed from its cage and placed immediately under the bell by raising one side a little. Within one minute the animal goes to sleep. As soon as its complete relaxation is observed, the animal is removed from the bell.

(c) The clinical thermometer, previously sterilized, is immediately inserted into the animal's rectum up to a mark which indicates the standard predetermined length of penetration: 4 cm for the rat and 3 cm for the jerboa. The thermometer is removed from the rectum $1\frac{1}{2}$ minutes after the penetration; at that time the animal begins to awaken. The time taken by the animal to awaken is ample to take its temperature with precision.

The difference of 1 cm in rectal penetration results from the jerboa's anatomy. The maximum length of penetration for taking the body temperature with precision was determined on the basis of preliminary tests on jerboas and rats.

TECHNIQUE

The reading is considered valid only when the animal is not agitated at the beginning of anesthesia. In general, it sleeps quietly and awakens in the same way. Otherwise, the taking of rectal temperature is postponed to another day.

INDIRECT CALORIMETRY

Equipment

The five apparatuses utilized in all the experiments of indirect calorimetry by means of confined air were described earlier in this appendix.

Method

1. The three animals to be studied are removed from their normal cages at 2100 h on the day before the experiment and placed in separate cages, without food or water.

2. The following day at 1400 h the air of the containers is evacuated through the water pump and pure air is introduced and brought to the temperature fixed for the experiment, with the containers' stoppers closed and the taps of the air-sampling tubes opened.

3. Thirty minutes after the temperature of the confined air is brought to the desired level, the animals are introduced into their respective containers, the stopper and their taps are firmly closed, the time of closure for each container is recorded and the experiment begins (generally about 3 p.m.). At the same time the atmospheric pressure, environmental temperature and relative humidity of the air in the laboratory are recorded.

4. In the course of the experiment the state of the animals in the containers is observed from time to time by means of a stethoscope applied to the rubber stoppers. Usually, the preliminary tests condition the animal to keep quiet. When the animal is agitated the result is not considered valid, and the experiment is repeated.

5. The duration of each experiment is dictated by the temperature so as not to allow a carbon dioxide accumulation of more than one per cent.

6. At the end of the experiment a sample of confined air is drawn from each container into its bottle and the exact time when the sample is taken is recorded. The method of sampling is as follows: The container's tap is connected to the upper end of the sampling bottle, the bottle is filled with mercury, and the container's ventilator is set in motion to mix the confined air. The first sample of air withdrawn from the container into the bottle is returned to the container by manoeuvring the mercury reservoir, and only the second sample of air taken into the bottle is kept. The taps are closed, and the experiment ends. This operation takes three minutes for each container. When the sampling of confined air is completed the animals are weighed and returned to their cages.

7. The following day the samples of confined air are analysed in the Carpenter apparatus according to Haldane's (1935) well-known method. The first transfer of air from the sampling bottle to the burette of the apparatus is returned to the bottle, and it is the second transfer that is analysed. From time to time, to ensure the exactitude of results, the same sample of air is analysed twice.

APPENDIX I

8. Before each series of experiments, preliminary tests are made during which the animals are habituated to the experimental method adopted. The experimental data are represented in the form of curves. Several experiments have been made on each animal in order to achieve greater precision.

The following form (an actual record of J47) was used to record and calculate the results of experiments in indirect calorimetry. The correction coefficients and factors utilized were those given by Carpenter (1948).

EXPERIMENTAL RECORD OF ENERGY METABOLISM

Experiment No.: 546 *Nature*: Energy metabolism of the jerboa
Date: 25.8.57 *Lab. temp.*: 27·5° C *Animal No.*: J47
Begin: 1437 h *Rel. hum.*: 61% *Sex*: Male
End: 1649 h *Atm. press.*: 761·75 mm Hg *Diet*: Wet
Duration: 2 h 12 min *Weight*: 165 g (2)
Factor: 2·2 (1)* *Fasting period*: 18 h
Degree of tranquillity: 100%

Container II
 Exptl. temp.: 30° C Air content: 35·701 l.
  Animal's vol.: 0·165 l.
  $\overline{}$
  35·536 l. (a)

To find volume of air at 0° C and 760 mm Hg (STP):
 Water vapour pressure saturation at 27·5° C (lab. temp.)
  = 27·56 mm Hg (Carp. T3)
 At relative humidity (in lab.) of 61%
  = 27·56 × 0·61 = 16·81 mm Hg
 Atmospheric pressure at 30° C and STP
  = 761·75 − 16·81 = 744·94 mm Hg
(Carp. T10) *Factor*: 0·883 (b)
Therefore volume of air at STP in container is
  35·536 (a) × 0·883 (b) = 31·378 l. (3)

Burette
 Volume of air = 10·020 cm³
 After abs. CO − 9·943 = 0·077

$$\frac{0.077 \times 100}{10.020} = 0.7684 \ \% \ CO_2 \ (4)$$

 After abs. O_2 − 7·940 = 2·003

$$\frac{2.003 \times 100}{10.020} = 19.9900 \ \% \ O_2 \ (5)$$

Air 100% − (19·9900 + 0·7684) = 79·2416 % N_2 (6)
Carpenter Formula (1948): (79·030: % N_2: 20·940: *x*)
 Carp. T37 20·996 CO_2 measured 0·768 (4)
 % O_2 (5) − 19·990 (5) CO_2 in air − 0·030
 O_2 consumed 1·006 (7) CO_2 excreted 0·738 (8)

Therefore, R.Q. is: $\dfrac{0.738 \ (8)}{1.006 \ (7)} = 0.733$ (9)

The calorie value of R.Q. 0·733 is 4·717 (10)
 (Carp. T13)

Calorification (calculations):

container's air volume × O_2 consumed = 31·378 (3) × 0·01006 (7) = 0·3156 l. (11)
kcalories per hour: 0·3156 (11) × 4·717 (10) = 1·4886/2·2 (1) = 0·6766 (12)
kcal/g/h: 0·6766 (12)/165 (2) = 0·0041006 (13)
kcal/kg/h: 0·0041006 (13) × 1000 = 4·1006 (14)
kcal/kg/24 h: 4·1006 × 24 = 98·4144 (15)
kcal/24 h: 0·6766 (12) × 24 = 16·2384 (16)
kcal/kg × $^{0·73}$/24 h: factor by Brody T13·8 = 0·2659
            16·2384 (16)/0·2659 = 61·0695 (17)
kcal/10 kg$^{2/3}$/24 h: factor by Brody T13·8 = 0·2975
            16·2384 × 10 = 162·384/0·2975 = 545·825 (18)
O_2 l./kg/h: 0·3156 (11)/2·2 (1) = 0·1434545/165 (2)
            0·0008694 × 1000 = 0·8694 (19)

Note: The numbers in parentheses indicate the results of calculations for cross-reference.

The above form is prepared on the basis of the method of indirect calorimetry established by Brody (1945).

HEAT DISSIPATION AND EVAPORATIVE WATER LOSS

Insensible perspiration

The method adopted for the study of water loss through cutaneous and pulmonary evaporation is based on the following theoretical consideration: An animal's insensible perspiration, during a definite period, is equal to the difference between its initial and final body weights plus the weight of oxygen consumed; from this sum the weight of carbon dioxide produced is deducted, provided the animal does not excrete feces and urine during the experiment. That is

$$P = W - w + O_2 - CO_2 \text{ (without excrements)}$$

where P is insensible perspiration; W the initial body weight in grams; w the final body weight in grams; O_2 the oxygen consumed in grams; and CO_2 the carbon dioxide excreted in grams.

All the experiments of the series on insensible perspiration were conducted on the basis of this formula. The experiment was considered valid only when the animal kept quiet and discharged no excrements.

Equipment: The equipment used for this series of experiments comprised the following:

1. The five units of apparatus of indirect calorimetry described previously,
2. Precision scales (accurate to one milligram),
3. The animal weighing-box (in transparent plastic material), 8 cm long, 8 cm wide and 8·5 cm high, bearing holes in the sides and in the cover for ventilation. The animal placed in this box learns to keep quiet during the process of weighing.

Method: The method used for the study of insensible perspiration is exactly the same as that described for indirect calorimetry (see above). Moreover, the weighing of the animal must be done with precision, immediately before and immediately after the experiment.

APPENDIX I

The recording of insensible perspiration experiments comprises two parts: (a) calculation of the animal's energy metabolism (see Exp. No. 546, above) and (b) measurement of insensible perspiration.

EXPERIMENTAL RECORD OF INSENSIBLE PERSPIRATION

Experiment No. 546 *Nature:* Insensible perspiration of jerboa J47

Weighing time	Body weight	Excreta
Begin: 1438 h	Initial: 165·030 g (1)	Feces = 0
End: 1712 h	Final: 164·725 g	Urine = 0
Duration: 2 h 34 min	Difference = 0·305 g (2)	
Factor: 2·57 (3)		

(a) Insensible weight loss in g/kg/h:

$$0.305\ (2)/2.57\ (3) = 0.118677/165.030\ (1) = 0.0007191 \times 1000$$
$$= 0.7191\ (4)$$

(b) O_2 consumed g/kg/h (see energy metabolism Expt. No. 546):

$$O_2\ \text{l./kg/h}\ 0.8694 \times 1.42904 = 1.2424\ (5)$$

(c) CO_2 excreted g/kg/h (see energy metabolism Expt. No. 546):

Container's air volume (l.):	31·375 (3)
CO_2 excreted %:	0·738 (8)
Expt. duration (h):	2·2 (1)
Animal's body weight:	165 g

Therefore, CO_2 l./kg/h (volume) is:

$$31.375 \times 0.00738 = 0.2315696/2.2 = 0.1052589/165 = 0.0006379 \times 1000 = 0.6379\ \text{l.}$$

and CO_2 g/kg/h (weight) is

$$0.6379 \times 1.9769 = 1.2610\ (6)$$

(d) Water loss through insensible perspiration (cutaneous and pulmonary evaporation):

CO_2 excreted (g):	1·2610	(6)
O_2 consumed (g):	−1·2424	(5)
Surplus of CO_2:	0·0186	(7)
Insensible weight loss (g/kg/h):	0·7191	(4)
Surplus of CO_2 (g):	−0·0186	(7)
Insensible perspiration (g/kg/h)	=0·7005	(8)

Note: O_2 density at 0° C and 760 mm Hg = 1·42904 g/l.
CO_2 density at 0° C and 760 mm Hg = 1·9769 g/l.

Heat balance and water balance

The following experiments were conducted in order to: (a) study the effect of diet on the gains and losses of heat and of water in jerboas and rats; (b) partition these gains and losses to their respective routes of realization; and (c) determine the heat balance and the water balance per 100 g of body weight per day.

Experiments: The animals studied were: two jerboas on wet diet, two jerboas on dry diet, two white rats on wet diet and two white rats on dry diet.

The animals lived for six days, each pair in a separate compartment of the metabolism cage.

The month of June was chosen for the six-day experiments, as the atmospheric conditions varied very little: mean environmental temperature, 25° C; relative humidity, 62 per cent; atmospheric pressure, 760·75 mm Hg.

Each day at 1500 h, the following operations were performed: the body weight of each animal and the quantity of food consumed by each pair of animals were determined; the feces of each pair of animals were collected for drying until the weight remained constant (at 105° C); the urine and funnel-washing water for each pair of animals were collected for filtering and drying to constant weight.

The water content of wheat and lettuce consumed in the diet was determined by desiccation to constant weight.

The mean values per pair of animals and per 100 g of body weight per day were calculated.

Heat gains: The animals' heat gains were calculated on the basis of food consumed per 100 g of body weight per day, as follows:

$$1 \text{ g wheat} = 3 \cdot 6 \text{ kcal. (Hawk, 1947, p. 1246)}$$
$$1 \text{ g lettuce} = 0 \cdot 18 \text{ kcal. (Hawk, 1947, p. 1242)}$$

Heat losses: The animals' heat losses were calculated as follows: The difference between the basal metabolism (after 18 hours of fasting) and the metabolism (without fasting) in the course of normal life was considered to represent the extra heat produced by normal activity. The calories that the animal stores in the form of reserves (increase in body weight), or loses from its reserves (decrease in body weight), were calculated from the difference between its total heat gains and its total heat losses.

Water balance: The animals' water intake and water output per 100 g of body weight per day were calculated on the basis of data derived from the above experiments.

EXPERIMENTAL RECORD OF NUTRITION AND EXCRETION

(Actual record of the experiment on jerboas J66 and J67, on wet diet, to determine heat balance and water balance)

1. *Body weight* — grams
   - Mean for the first day (per animal, 297/2) — 148·5
   - Mean for six days (per animal, 1819·5/12) — 151·6
   - Mean increase of weight (per animal per day, 151·6–148·5) — 3·1
   - Mean increase of weight (per 100 g of body weight per day, 3·1 × 100/151·6) — 2·044

2. *Food*
   - Wheat: Consumption per animal per day (108/12) — 9·0
     - Consumption per 100 g of body weight per day (9 × 100/151·6) — 5·96
   - Lettuce: Consumption per animal per day (89/12) — 7·416
     - Consumption per 100 g of body weight per day (7·416 × 100/151·6) — 4·892
   - Natural water content
     - Wheat water: 11·7%. Consumption per 100 g body weight (5·96 × 11·7/100) — 0·697
     - Lettuce water: 66%. Consumption per 100 g body weight (4·892 × 66/100) — 3·228
   - Water of metabolism derived from wheat consumed per 100 g body weight per day:
     - 5·96 × 0·13 (prot.) = 0·7748 × 0·41 (H_2O) = 0·318
     - 5·96 × 0·02 (fat) = 0·1192 × 1·07 (H_2O) = 0·127
     - 5·96 × 0·724 (carb.) = 4·315 × 0·6 (H_2O) = 2·589
     - — 3·034
   - From lettuce consumed per 100 g body weight per day:
     - 4·892 × 0·012 = 0·0587 × 0·41 = 0·024
     - 4·892 × 0·002 = 0·00978 × 1·07 = 0·010
     - 4·892 × 0·029 = 0·14186 × 0·6 = 0·085
     - — 0·119

APPENDIX I

3. *Excretion*

    Feces: Fresh weight 1·270 g
            Dry weight   0·675 g

Water 0·595 g	
Dry residue per animal per day (9·760/12)	0·813
Fresh feces per animal per day	
(1·270 × 9·760/0·675 = 18·363/12)	1·530
Water loss per animal per day (1·530−0·813)	0·717
Fresh feces per 100 g body weight per day	
(1·530 × 100/151·6)	1·009
Dry residue per 100 g body weight per day	
(0·813 × 100/151·6)	0·536
Water loss per 100 g body weight per day (1·009−0·536)	0·473
Percentage of fresh weight (0·473 × 100/1·009)	46·8%
Percentage of dry weight (0·473 × 100/0·536)	88·2%

    Urine: Fresh weight 7·000 g
           Dry weight   0·287 g

Water 6·713 g	
Dry residue per animal per day (2·200/12)	0·183
Fresh urine per animal per day	
(2·200 × 7/0·287 = 53·658/12)	4·471
Urine water per animal per day (4·471−0·183)	4·288
Fresh urine per 100 g body weight per day	
(4·471 × 100/151·6)	2·949
Dry residue per 100 g body weight per day	
(0·183 × 100/151·6)	0·121
Urine water loss per 100 g body weight per day (2·949−0·121)	2·828
Percentage of fresh weight (2·828 × 100/2·949)	95·9%
Percentage of dry weight (2·828 × 100/0·121)	2337%

Water and fat contents of the animal body

It is known that the fatty tissues of the animal body contain small amounts of water (7–20 per cent) compared with other tissues. Therefore, in order to determine the water content of the animal body, it is preferable to take as a basis not the weight of the entire body but the weight of the body without its fatty tissues.

Accordingly, two series of experiments were conducted: (a) to determine the weight of the dry residue of the animal body; and (b) to determine the weight of the total fat contained in the dry residue.

By deducting the weight of the animal's fat from its total body weight, the water content can be expressed as a percentage of the fat-free body weight according to the formula

$$W = f + r + w$$

where W is the body weight; f the weight of total fat; r the dry weight of fat-free residue; and w the weight of water. To determine these values, we used the material and method described below.

Equipment: Meat-mincing machine; containers to include the entire minced body, with the water used for washing; oven to dry at 105° C; scales of 1-mg precision.

Method: Allow the animals to fast for 18 hours; weigh each animal and anesthetize; mince each animal separately and place the mince, with the water used for washing, in a container of known weight; place the container in the oven at 105° C; weigh periodically, until it reaches a constant weight; pulverize the residue and again place it in the oven until it reaches a constant weight.

To determine the weight of fat in the dry residue, the test recommended by Gradwohl (1948) was used: Place 0·5 g of dry residue in a graduated tube, containing 10 cm³ of 30 per cent hydrochloric acid. Boil it in a waterbath for 15 minutes. After cooling, add ether up to the 50-cm³ mark, and stopper the opening of the tube securely. Invert the tube 40–50 times and allow it to stand for half an hour. With a pipette withdraw 20 cm³ of the supernatant fluid and put into a drying flask of known weight. Evaporate and dry the residue in a waterbath. Cool and weigh.

EXAMPLE OF CALCULATION: RAT 27

	grams
Initial body weight	224·5
Dry residue of the body	− 78·43
Water	146·07
Fat content per gram of dry residue	0·340
Fat content (78·43 × 0·340)	26·666
Therefore, Rat 27 is composed of:	
Fat	26·666
Dry residue (78·43 − 26·666)	51·764
Water	146·070
Total body weight,	224·500
Body weight without fat (224·5 − 26·666)	187·834
Percentage of water contained in the body, without fat:	
146·07 × 100/187·834 =	77·7%

ACTOGRAPH

To make a comparative study of the 24-hour activity cycle of animals, we constructed an actograph (*Figure 29*), which registers the tracings of the

Figure 29. Actograph.

APPENDIX I

normal activity of three animals at the same time. Our actograph is constituted of a kymograph and three cages conveniently suspended, as follows:

1. The kymograph is a wooden cylinder 23·4 cm in radius, 147 cm in circumference and 40 cm in height. Its axis is connected to the axis of a clock which makes a complete revolution every 24 hours and 30 minutes. The kymograph therefore moves at a linear speed of 1 mm/min.

2. The three cages of the actograph are separated, and each cage rests on the two fixed points of one side and on the airtight bellows of the opposite side. Rubber tubing connects the bellows to the Marey receiving tambours, which inscribe on the kymograph the slightest movement of each cage: cage movements cause the elastic tissue of the bellows and tambours to vibrate by the compression and decompression of confined air. The disposition of the Marey drums and of the pens of the three cages is such that they inscribe simultaneously. Four tracings of the same animal can be made, one below the other, during four consecutive days.

Appendix 2

BODY WEIGHT OF JERBOAS

Jerboas grouped according to body weight

Males

No.	Weight (g)	Percentage of the total by zone of weight	No.	Weight (g)	Percentage of the total by zone of weight
J33	190		J47	161	
J70	186	Group I	J46	160	Group II
J14	185		J67	160	(continuation)
J31	182	28%	J66	157	
J38	182		J29	156	60%
J43	182		J49	156	
J32	180		J27	153	
			J74	152	
J71	179		J73	151	
J25	173	Group II			
J62	173		J30	146	
J69	173		J42	144	Group III
J53	165		J51	144	
J19	164				12%

Average of 25 males = 166 g   Range = 144–190 g

Females

No.	Weight (g)	Percentage of the total by zone of weight	No.	Weight (g)	Percentage of the total by zone of weight
J63	190	Group I	J36	148	Group III
J37	188		J76	147	
J72	186	16%	J18	147	40%
J34	180		J22	146	
			J44	145	
			J48	145	
J12	175		J77	145	
J40	175	Group II	J24	145	
J20	174		J21	144	
J16	165	44%	J28	141	
J39	164				
J45	163				
J41	162				
J68	162				
J50	157				
J35	155				
J15	153				

Average of 25 females = 160 g   Range = 141–190 g

APPENDIX 3

DIMENSIONS OF THE JERBOA

J70, male

Description	Measurement
Body weight	186 g
Body lengths (animal in extension on its back):	
From nose to anus	160 mm
From origin of sternum to first caudal vertebra	110 mm
From extremity of sternum to first caudal vertebra	80 mm
Average length of tail	250 mm
Cranium	
Maximum length	40 mm
Maximum width	30 mm
Diameter of eyeball	8 mm
Ears	
Height	35 mm
Width	25 mm
Trunk girth measured at the level of	
Axilla	85 mm
Extremity of sternum	95 mm
Abdomen	120 mm
Forelimbs	
Humerus (arm)	15 mm
Cubitus-radius (forearm)	20 mm
Carpus-metacarpus	5 mm
Length of fingers	5 mm
Length of nails	5 mm
Hind-limbs	
Femur	35 mm
Tibia	70 mm
Tarsal-metatarsal bone (one)	45 mm
Length of toes (three)	20 mm
Length of nails	5 mm

Appendix 4

TEMPERATURE AND HUMIDITY OF THE JERBOA'S HABITAT

Temperatures inside and out of the jerboa's burrow

Time (h)	Outside					Inside		
	Temperature (°C)			Water (mg/l.) of air	Relative humidity (%)	Temperature (°C)	Water (mg/l.) of air	Relative humidity (%)
	Sun	Soil	Shade					
1000	30·5	37·0	28·5	20·7	64	25·1	22·3	75
1100	30·5	37·5	28·9	20·7	63	25·7	23·1	75
1200	31·0	39·0	29·4	20·3	60	25·7	24·2	78
1300	31·5	40·0	29·4	20·3	60	26·8	26·2	80
1400	31·0	39·0	28·9	20·7	63	26·8	26·2	80
1500	30·0	38·5	28·9	20·7	63	26·8	26·2	80
1600	29·5	35·5	27·8	19·9	65	25·7	24·5	79
1700	28·0	32·0	28·2	20·5	69	25·7	23·8	77

Appendix 5

COMPARATIVE GROWTH OF THE JERBOA AND RAT

Jerboas

Age in days	Mean weight* (g)	Observations on development
1		Born naked, without hair; ears and eyes closed
6	7·5	
18	16·5	Ears open
20	—	Hair appears
27	24·0	
30	—	Dentition
34	25·5	
36	—	Eyes open
		Forelimbs develop
39	28·5	
46	34·0	
53	43·0	Hind-limbs develop
60	53·5	Eat solid foods
67	66·5	
74	79·5	
81	82·5	Jump
95	88·5	
102	93·0	(Sexual maturity: Two spiny prongs on the back of the head of the penis in the male)
117	103·0	
135	116·0	
144	121·5	
158	130·0	
174	135·5	
193	145·5	
236	146·0	

* Mean weight of animals whose growth curve is established.

(continued)

COMPARATIVE GROWTH OF THE JERBOA AND RAT

Appendix 5 (*continued*)

White rats

Age in days	Mean weight* (g)	Observations on development
1		Born naked, without hair; ears and eyes closed
7	10·5	Ears open
10	—	Hair appears
13	—	Eyes open
15	—	Dentition. Forelimbs and hind-limbs develop
16	15·5	
21	—	Eat solid foods
23	22·5	
28	27·5	
35	37·0	
42	48·0	
49	53·0	
56	63·0	
60	—	Sexual maturity
63	75·0	
70	86·0	
77	94·5	
84	97·0	
91	109·0	
100	114·0	
106	119·0	
124	132·0	
133	137·0	
147	144·0	
162	146·0	
181	146·5	
212	148·0	

* Mean weight of animals whose growth curve is established.

Appendix 6

VARIATION OF THE JERBOA'S BODY WEIGHT DURING GROWTH AND AFTER MATURITY

Age in days	Male jerboas			Mean weight* (g)	Female jerboas			Mean weight* (g)
	J47	J49	J43		J50	J45	J44	
81	81	81	90	84	81	90	83	84
88	86	87	98	90	76	98	85	86
95	94	91	111	98	83	105	92	83
104	104	95	117	105	99	117	98	104
121	118	118	128	121	115	126	105	115
126	126	122	144	131	126	143	120	129
137	141	132	150	141	139	146	116	134
146	147	140	155	147	142	146	117	135
150	153	148	165	155	147	147	124	139
157	155	153	172	160	157	154	132	147
164	161	156	175	164	161	156	137	151
172	156	156	180	164	161	163	143	156
181	158	159	182	166	153	155	145	151
191	155	159	182	165	157	158	150	155
199	160	157	170	162	151	153	146	150
209	152	159	172	161	152	160	150	154
218	153	156	170	159	155	158	151	154
229	143	—	171	157	148	158	150	152
235	—	139	166	152	—	153	149	151
250	142	—	164	153	147	154	150	150
264	148	145	165	153	150	153	145	149
274	151	152	164	155	143	157	139	146
283	150	148	166	154	143	153	138	144
296	150	153	167	156	135	157	138	143
311	150	137	171	153	150	159	145	151
320	148	143	157	149	—	150	134	142
349	150	150	158	152	143	146	136	141
363	152	147	164	154	151	150	—	150
375	156	155	167	159	141	150	—	145

* Mean weight of animals whose growth curve is established.

Appendix 7

EFFECT OF DRY DIET

Effect of dry diet on body weight of adult jerboas

Date	Age in days	Diet	Body weight (g)	
			J46 Male	J41 Female
17.06.54	81	Wet	81	75
24.06.54	88		86	77
01.07.54	95		90	81
10.07.54	104		95	90
27.07.54	121		105	110
01.08.54	126		—	117
12.08.54	137		124	133
21.08.54	146		127	142
25.08.54	150		132	151
01.09.54	157		143	157
08.09.54	164		148	162
16.09.54	172		159	158
25.09.54	181		160	160
05.10.54	191	Dry	157	156
13.10.54	199		159	158
23.10.54	209		161	165
01.11.54	218		165	164
12.11.54	229		166	162
18.11.54	235		168	160
03.12.54	250		163	160
17.12.54	264		160	160
27.12.54	274		165	161
05.01.55	283		172	161
18.01.55	296		173	161

(*continued*)

APPENDIX 7

Body-weight variations of jerboas on dry diet

Date	J35 Female (g)	J19 Male (g)	J22 Female (g)	J41 Female (g)	Date	J51 Male (g)
09.02.54	150	168			27.12.54	149
13.02.54	153	156				
20.02.54	155	144			05.01.55	152
15.03.54	156	148			18.01.55	157
04.04.54	151	150			02.02.55	147
23.05.54	158	149			12.03.55	141
05.06.54	165	162	150		07.04.55	146
01.07.54	157	157	150		18.05.55	153
27.07.54	162	165	143		07.06.55	152
12.08.54	167	170	138		27.07.55	167
25.09.54	151	175	156	160	10.09.55	152
05.10.54	155	175	149	156	10.10.55	144
23.10.54	172	178	149	165	02.11.55	152
18.11.54	162	162	148	160	06.12.55	157
27.12.54	165	148	154	158	01.02.56	156
05.01.55	168	155	155	161	05.03.56	155
18.01.55	171	150	150	—	20.05.56	158
02.02.55	160	140	157	169	27.07.56	160
12.03.55	155	142	155	169	27.08.56	156
07.04.55	153	145	156	174	17.09.56	154
18.05.55	153	164	162	170	19.10.56	152
07.06.55	155	147	151	176	01.12.56	153
28.06.55	—	106	—	177	17.01.57	140
27.07.55	160	Inanition	149	180	08.08.57	153
10.09.55	—		152	153		
10.10.55	157		147	97	29.03.58	143
02.11.55	150		154	Inanition	12.05.58	141
06.12.55	145		139			
01.02.56	149		130			
24.04.56	137		Inanition			
Duration of resistance to dry diet	2 y 2 m 15 d	1 y 4 m 17 d	1 y 7 m 26 d	1 y 15 d		3 y 5 m 16 d

Appendix 8

EFFECT OF WET AND DRY DIET

Effect of alternating wet and dry diets on body weight of jerboas

Date	J38, *male*		Date	J63, *male*	
	Body weight (g)	Diet		Body weight (g)	Diet
09.02.54	190	Dry	18.03.55	136	Wet
13.02.54	182		07.06.55	132	
20.02.54	172		27.07.55	134	
15.03.54	164		10.09.55	132	
04.04.54	173		21.10.55	143	
23.05.54	175		06.12.55	153	
05.06.54	185		01.02.56	164	
01.07.54	179		24.02.56	—	Dry
27.07.54	167		09.05.56	147	
12.08.54	172		27.07.56	175	
25.09.54	192		27.08.56	157	
23.10.54	193		17.09.56	154	
18.11.54	178		29.09.56	147	
03.12.54	120		09.10.56	140	
17.12.54	99		11.12.56	155	
17.12.54	Inanition	Wet	11.12.56	—	Wet
05.01.55	148		08.08.57	150	
18.01.55	172		21.03.58	172	
02.02.55	180		12.05.58	180	
12.03.55	181		09.06.58	180	
07.04.55	185		27.10.58	190	
18.05.55	172		22.11.58	196	
07.06.55	180				
27.07.55	203				
10.09.55	201				

(continued)

APPENDIX 8

Effect of wet and dry diets on the excretion of jerboas

Group I

Jerboas habituated to wet diet for a period of more than one year
Mean ambient temperature: 26–27° C

Animal	Date	R.H. (%)	Consumption		Body weight (g)	Feces (g)	Urine (cm^3)	Urine reaction (pH)
			Wheat (g)	Lettuce (g)				
J14, male	Wet diet							
	18.8.53	60	11·0	9	181	0·94	3·0	8
	19.8.53	65	12·5	9	183	1·32	2·0	8
	20.8.53	67	12·5	8	185	1·35	4·3	8
	22.8.53	66	11·5	9	187	1·45	4·8	8
	23.8.53	65	12·0	9	188	1·36	2·7	7
	24.8.53	70	13·0	10	189	1·31	3·1	7
	Mean		12·08	9	185·5	1·28	3·31	8
	Dry diet							
	25.8.53	61	12·0		190	1·20	1·1	5
	26.8.53	58	9·0		189	0·80	0·8	5
	27.8.53	61	11·0		188	1·07	1·0	5
	28.8.53	68	7·0		189	0·87	0·5	5
	29.8.53	66	5·5		185	0·51	1·2	5
	30.8.53	63	8·5		185	0·61	0·8	5
	Mean		8·83		187·6	0·84	0·90	5
J25, male	Wet diet							
	31.8.53	65	10·0	4	168	0·96	2·9	7
	01.9.53	70	10·5	7	171	1·10	5·0	7
	02.9.53	68	11·0	3	172	1·15	4·0	7
	03.9.53	71	10·5	5	172	1·46	4·0	6
	04.9.53	73	11·0	10	172	1·00	4·8	7
	05.9.53	67	12·0	8	174	1·37	3·4	7
	Mean		10·83	6·16	171·5	1·17	4·01	7
	Dry diet							
	06.9.53	63	11·0		174	1·14	2·9	6
	07.9.53	62	11·0		173	1·12	0·5	4
	08.9.53	48	7·5		173	0·90	1·8	4
	09.9.53	51	8·5		170	0·51	1·0	4
	10.9.53	57	9·0		169	0·91	1·0	4
	11.9.53	58	9·5		167	0·81	0·5	4
	Mean		9·41		171	0·89	1·28	4

EFFECT OF WET AND DRY DIET

Animal	Date	R.H. (%)	Consumption		Body weight (g)	Feces (g)	Urine (cm³)	Urine reaction (pH)
			Wheat (g)	Lettuce (g)				
J29, male	Wet diet							
	31.8.53	65	9·0	3	143	0·94	2·3	8
	01.9.53	70	9·0	10	150	0·99	6·0	7
	02.9.53	68	9·0	6	150	1·24	3·1	7
	03.9.53	71	9·0	7	150	1·05	3·3	7
	04.9.53	73	11·0	10	152	1·22	3·1	8
	05.9.53	67	9·5	7	155	1·15	2·8	7
	Mean		9·41	7·16	150	1·09	3·44	7
	Dry diet							
	06.9.53	63	9·5		156	1·04	2·5	5
	07.9.53	62	9·0		153	0·87	1·5	4
	08.9.53	48	8·3		152	0·72	0·1	5
	09.9.53	51	7·5		151·5	0·66	1·0	5
	10.9.53	57	9·5		151	0·62	1·3	5
	11.9.53	58	8·5		149	0·76	0·2	4
	Mean		8·75		152·1	0·78	1·1	5
J19, male	Wet diet							
	12.9.53	66	9·5	8	186	1·16	2·7	8
	13.9.53	68	10·5	9	188	1·20	3·0	6
	14.9.53	68	10·0	4	189	1·10	3·2	6
	15.9.53	62	10·0	8	185	1·05	2·7	8
	16.9.53	59	11·5	8	187	1·32	2·8	8
	17.9.53	62	11·0	6	188	1·45	2·8	8
	Mean		10·41	7·16	187·16	1·21	2·86	8
	Dry diet							
	18.9.53	67	10·5		187	0·99	1·5	5
	19.9.53	68	10·0		185	0·86	1·2	5
	20.9.53	64	10·5		183	0·74	0·3	5
	21.9.53	64	10·5		183	0·95	0·2	5
	22.9.53	63	8·5		182	0·76	0·6	5
	23.9.53	67	8·5		180	0·90	0·3	5
	Mean		9·75		183·33	0·86	0·68	5
J3, female	Wet diet							
	18.8.53	60	6	9	152	0·85	3·0	8
	19.8.53	65	6	9	132	1·07	2·0	8
	20.8.53	67	7	9	132	0·62	3·0	8
	22.8.53	66	8	9	135	0·75	2·0	8
	23.8.53	65	8	5	136	0·65	1·5	7
	24.8.53	70	8	8	136	0·90	2·7	7
	Mean		7·16	8·16	133·83	0·80	2·36	8

APPENDIX 8

Animal	Date	R.H. (%)	Consumption		Body weight (g)	Feces (g)	Urine (cm³)	Urine reaction (pH)
			Wheat (g)	Lettuce (g)				
J3, female *(continued)*	Dry diet							
	25.8.53	61	8.0		135	0.55	1.2	5
	26.8.53	58	6.0		135	0.62	0.3	4
	27.8.53	61	6.5		135	0.45	0.2	5
	28.8.53	66	6.0		134	0.65	0.3	5
	29.8.53	66	6.0		134	0.30	0.3	5
	30.8.53	63	7.0		135	0.36	0.3	5
	Mean		6.58		134.66	0.48	0.43	5
J12, female	Wet diet							
	18.8.53	60	12.0	10	182	0.94	3.5	9
	19.8.53	65	11.5	10	183	1.35	3.5	8
	20.8.53	67	10.0	10	183	1.60	5.2	8
	22.8.53	66	9.0	10	182	1.00	5.5	8
	23.8.53	65	8.0	10	182	1.26	4.4	7
	24.8.53	70	6.5	10	181	0.25	6.5	8
	Mean		9.50	10	182.16	1.19	4.76	8
	Dry diet							
	25.8.53	61	6.0		176	0.82	1.2	4
	26.8.53	58	6.0		176	0.60	0.3	5
	27.8.53	61	3.5		175	0.66	0.2	5
	28.8.53	66	5.5		172	0.22	0.2	5
	29.8.53	66	5.0		170	0.52	0.1	4
	30.8.53	63	6.5		170	0.45	0.1	5
	Mean		5.41		173.16	0.54	0.35	5
J16, female	Wet diet							
	31.8.53	65	7.0	8	162	0.86	3.9	7
	01.9.53	70	8.0	9	164	0.71	5.3	8
	02.9.53	68	8.5	7	164	0.99	4.0	7
	03.9.53	71	7.5	5	165	1.20	3.5	7
	04.9.53	73	9.0	8	165	0.82	4.2	8
	05.9.53	67	8.0	8	165	0.98	3.7	7
	Mean		8.0	7.5	164.1	0.92	4.1	7
	Dry diet							
	06.9.53	63	9.0		166.0	0.77	1.8	3
	07.9.53	62	10.0		166.0	0.81	1.1	4
	08.9.53	48	9.0		165.5	0.85	1.1	5
	09.9.53	51	10.0		166.0	0.85	1.4	4
	10.9.53	57	10.0		166.5	0.85	1.0	4
	11.9.53	58	9.5		167.0	0.89	1.1	4
	Mean		9.58		166.16	0.83	1.25	4

EFFECT OF WET AND DRY DIET

Animal	Date	R.H. (%)	Consumption		Body weight (g)	Feces (g)	Urine (cm³)	Urine reaction (pH)
			Wheat (g)	Lettuce (g)				
J22, female	Wet diet							
	31.8.53	65	6	2	137·0	0·46	1·8	9
	01.9.53	70	7	7	140·0	0·75	3·9	8
	02.9.53	68	6	7	141·0	0·55	4·0	9
	03.9.53	71	7	2	140·5	1·06	2·0	7
	04.9.53	73	6	3	139·0	0·37	1·2	7
	05.9.53	67	8	2	139·0	0·61	1·8	8
	Mean		6·66	3·83	139·41	0·63	2·45	8
	Dry diet							
	06.9.53	63	9·5		140·0	0·96	1·5	5
	07.9.53	62	9·0		139·0	0·72	0·5	4
	08.9.53	48	5·5		137·0	0·43	0·6	5
	09.9.53	51	6·0		137·0	0·31	0·3	5
	10.9.53	57	8·5		136·5	0·57	0·7	4
	11.9.53	58	9·0		137·5	0·71	0·1	5
	Mean		7·91		137·66	0·61	0·61	5
J28, female	Wet diet							
	31.8.53	65	7·0	5	130·0	0·96	1·1	8
	01.9.53	70	9·0	6	134·0	2·58	2·6	8
	02.9.53	68	9·0	7	135·0	2·33	2·0	9
	03.9.53	71	9·0	7	135·0	1·91	2·1	7
	04.9.53	73	9·0	4	136·0	1·60	2·0	8
	05.9.53	67	3·5	9	135·5	1·37	3·2	8
	Mean		8·58	6·33	134·2	1·79	2·16	8
	Dry diet							
	06.9.53	63	9·0		135·0	1·04	1·2	6
	07.9.53	62	8·5		134·0	1·15	0·5	5
	08.9.53	48	7·5		133·5	1·11	0·1	4
	09.9.53	51	8·0		132·0	1·11	0·1	5
	10.9.53	57	8·0		131·5	1·03	0·1	5
	11.9.53	58	7·5		131·0	1·17	0·5	5
	Mean		8·08		132·8	1·10	0·41	5
J20, female	Wet diet							
	12.9.53	66	9·0	8	148	0·84	3·1	9
	13.9.53	68	9·5	8	151	0·91	4·2	8
	14.9.53	68	8·0	8	152	1·11	4·1	8
	15.9.53	62	9·0	8	150	1·06	4·3	9
	16.9.53	59	9·0	6	150	0·95	3·3	8
	17.9.53	62	8·5	7	150	1·14	4·3	8
	Mean		8·83	7·50	150·16	1·0	3·88	8

APPENDIX 8

Animal	Date	R.H. (%)	Consumption Wheat (g)	Consumption Lettuce (g)	Body weight (g)	Feces (g)	Urine (cm^3)	Urine reaction (pH)
J20, female *(continued)*	Dry diet							
	18.9.53	67	10.0		150.0	0.70	1.1	5
	19.9.53	68	9.0		149.5	0.86	0.6	5
	20.9.53	64	9.5		149.0	0.84	0.6	5
	21.9.53	64	9.0		148.5	1.15	0.7	5
	22.9.53	63	8.0		148.0	0.81	0.3	5
	23.9.53	67	8.0		147.0	0.87	0.2	4
	Mean		8.91		148.66	0.87	0.56	5
J21, female	Wet diet							
	12.9.53	66	8.5	8	137	0.61	3.3	9
	13.9.53	68	7.0	8	141	0.99	3.7	8
	14.9.53	68	8.0	7	141	0.98	2.4	8
	15.9.53	62	8.5	8	140	1.02	3.0	8
	16.9.53	59	8.5	6	141	1.03	3.2	8
	17.9.53	62	9.0	7	141	1.44	3.3	8
	Mean		8.25	7	140.16	1.01	3.15	8
	Dry diet							
	18.9.53	67	9.0		141	0.62	1.4	6
	19.9.53	68	9.5		140	0.82	0.5	5
	20.9.53	64	10.0		140	0.86	0.9	5
	21.9.53	64	10.5		140	0.92	0.2	5
	22.9.53	63	9.5		140	0.90	0.2	5
	23.9.53	67	9.5		139	1.01	0.1	5
	Mean		9.66		140	0.85	0.55	5
J27, male*	Wet diet							
	12.9.53	66	7.0	0	146	0.38	1.6	5
	13.9.53	68	8.0	0	146	0.63	2.1	5
	14.9.53	68	8.0	0	145	0.89	1.1	5
	15.9.53	62	9.0	0	145	0.75	1.7	5
	16.9.53	59	8.5	0	144	0.91	1.1	5
	17.9.53	62	10.5	0	144	0.82	1.5	5
	Mean		8.50	0	145	0.73	1.51	5
	Dry diet							
	18.9.53	67	10.5		144.0	0.86	1.2	5
	19.9.53	68	11.0		145.0	0.85	1.6	5
	20.9.53	64	12.0		145.0	0.99	1.6	5
	21.9.53	64	11.5		145.5	1.13	1.3	5
	22.9.53	63	11.5		146.0	1.13	1.1	5
	23.9.53	67	11.0		146.5	1.13	1.2	5
	Mean		11.25		145.3	1.01	1.33	5

* Exceptional case of a jerboa which refused to eat lettuce and followed dry diet at will.

EFFECT OF WET AND DRY DIET

Summary of Group I

(Mean values for six days)

Animal	Date	R.H. (%)	Consumption		Body weight (g)	Feces (g)	Urine (cm³)	Urine reaction (pH)
			Wheat (g)	Lettuce (g)				
Males	Wet diet							
J14	18.8.53	70–60	12·1	9·00	185·5	1·28	3·31	8
J25	31.8.53	73–65	10·8	6·16	171·5	1·17	4·01	7
J29	31.8.53	73–65	9·4	7·16	150·0	1·09	3·44	7
J19	12.9.53	68–59	10·4	7·16	187·2	1·21	2·86	8
Mean		71–62	10·6	7·37	173·5	1·19	3·40	8
Females								
J3	18.8.53	70–60	7·2	8·16	133·8	0·80	2·36	8
J12	18.8.53	70–60	9·5	10·00	182·2	1·19	4·76	8
J16	31.8.53	73–65	8·0	7·50	164·1	0·92	4·10	7
J22	31.8.53	73–65	6·7	3·83	139·4	0·63	2·45	8
J28	31.8.53	73–65	8·6	6·33	134·2	1·79	2·16	8
J20	12.9.53	68–59	8·8	7·50	150·2	1·00	3·88	8
J21	12.9.53	68–59	8·2	7·00	140·2	1·01	3·15	8
Mean		70–61	8·1	7·19	149·1	1·05	3·26	8
Mean: M. & F.		70–62	9·0	7·25	158·0	1·10	3·32	8
Per 100 g body weight			5·7	4·58		0·69	2·10	
Males	Dry diet							
J14	25.8.53	68–58	8·8		187·6	0·84	0·90	5
J25	06.9.53	63–48	9·4		171·0	0·89	1·28	4
J29	06.9.53	63–48	8·7		152·1	0·78	1·10	5
J19	18.9.53	68–63	9·7		183·3	0·86	0·68	5
Mean		65–53	8·9		173·5	0·84	0·94	5
Females								
J3	25.8.53	66–58	6·6		134·7	0·48	0·43	5
J12	25.8.53	66–58	5·4		173·2	0·54	0·35	5
J16	06.9.53	63–48	9·6		166·2	0·83	1·25	4
J22	06.9.53	63–48	7·9		137·7	0·61	0·61	5
J28	06.9.53	63–48	8·1		132·8	1·10	0·41	5
J20	18.9.53	68–63	8·9		148·7	0·87	0·56	5
J21	18.9.53	68–63	9·7		140·0	0·85	0·55	5
Mean		65–55	8·0		147·6	0·75	0·59	5
Mean: M. & F.		65–54	8·35		157·1	0·79	0·74	5
Per 100 g body weight			5·31			0·50	0·47	

APPENDIX 8

Group II

Jerboas habituated to dry diet for a period of more than one year

Animal	Date	R.H. (%)	Consumption of wheat (g)	Body weight (g)	Feces (g)	Urine (cm³)	Urine reaction (pH)
J15, female	12.9.53	66	6·0	135	0·39	0·1	3
	13.9.53	68	7·0	136	1·30	0·2	3
	14.9.53	68	8·0	136	1·78	0·1	3
	15.9.53	62	8·5	136	1·04	0·2	3
	16.9.53	59	8·5	137	1·71	0·1	3
	17.9.53	62	8·5	137	1·79	0·2	3
	Mean		7·75	136·1	1·33	0·15	3
	18.9.53	67	8·0	137	0·78	0·5	3
	19.9.53	68	10·5	138	0·80	0·2	3
	20.9.53	64	12·5	140	1·14	0·7	3
	21.9.53	64	13·5	143	1·42	1·1	3
	22.9.53	63	13·5	145	1·66	1·0	4
	23.9.53	67	12·0	146	1·17	2·0	4
	Mean		11·16	141·5	1·16	0·91	3
J18, female	12.9.53	66	6·0	138	0·31	0·3	3
	13.9.53	68	5·0	138	0·27	0·2	3
	14.9.53	68	6·0	138	0·32	0·2	3
	15.9.53	62	6·5	139	0·55	0·3	3
	16.9.53	59	7·0	139	0·41	0·2	3
	17.9.53	62	5·5	139·5	0·58	0·3	3
	Mean		6·0	138·5	0·40	0·25	3
	18.9.53	67	7·0	139	0·55	0·2	3
	19.9.53	68	8·0	139	0·65	0·8	3
	20.9.53	64	8·0	139	0·63	0·3	3
	21.9.53	64	8·5	139·5	0·80	0·4	4
	22.9.53	63	7·0	141	0·83	0·7	4
	23.9.53	67	7·0	140	0·71	0·2	2
	Mean		7·58	139·5	0·69	0·43	3

(continued)

EFFECT OF WET AND DRY DIET

Summary of Group II

(Mean values for twelve days)

Animal	Date	R.H. (%)	Consumption of wheat (g)	Body weight (g)	Feces (g)	Urine (cm³)	Urine reaction (pH)
Females							
J15	12.9.53	68–59	7·75	136·16	1·33	0·15	3
J15	18.9.53	68–63	11·66	141·50	1·16	0·91	3
J18	12.9.53	68–59	6·00	138·58	0·40	0·25	3
J18	18.9.53	68–63	7·58	139·58	0·69	0·43	3
Mean		68–61	8·24	138·95	0·89	0·43	3
Per 100 g body weight			5·92		0·64	0·31	

(*continued*)

APPENDIX 8

Effect of temporary wet and dry diets on excretion of jerboas and white rats

(See Appendix 1)

Duration of the experiments   11.6.1958–16.6.1958
Mean environmental temperature   25° C
Mean environmental relative humidity   62 per cent
Mean atmospheric pressure   760·73 mm Hg

Description	Wet diet		Dry diet	
	Jerboas J66, J67 (g)	Rats R32, R33 (g)	Jerboas J42, J51 (g)	Rats R31, R22 (g)
1. Body weight				
Mean per animal:				
First day	148·5	161·5	147	214·8
Six days	151·6	166	147	206
Difference	3·1	4·5	0	−8·8
Difference per 100 g	2·04	2·710	0	−4·2
2. Food ingested per day				
Wheat: per animal	9·0	10·6	5·7	1·46
per 100 g body weight	5·96	6·38	3·87	0·70
Lettuce: per animal	7·41	27·16	0	0
per 100 g body weight	4·89	16·36	0	0
Natural water contained*				
Wheat	0·69	0·74	0·45	0·09
Lettuce	3·22	7·52	0	0
Water of metabolism*				
Wheat	3·04	3·25	1·97	0·36
Lettuce	0·11	0·40	0	0
3. Excretion				
Feces				
Sample, fresh weight	1·27	3·45	1·37	0·33
dry weight	0·68	1·34	0·80	0·19
Water eliminated by feces	0·59	2·11	0·57	0·14
Per day:				
Total dry residue per animal	0·81	1·65	0·52	0·54
Fresh feces per animal	1·53	4·25	0·89	0·92
Water loss per animal	0·72	2·60	0·37	0·38
Fresh feces*	1·00	2·56	0·60	0·44
Dry residue*	0·53	0·99	0·35	0·26
Water loss*	0·47	1·57	0·25	0·18
Water content of feces:				
% of fresh weight	46·8	61·2	41·7	41·2
% of dry weight	88·2	157·7	71·7	70·3
Urine				
Sample, fresh weight	7·00	10·00	1·40	3·00
dry weight	0·28	0·33	0·14	0·26
Water eliminated by urine	6·72	9·67	1·26	2·74

* per 100 g body weight per day

EFFECT OF WET AND DRY DIET

Description	Wet diet		Dry diet	
	Jerboas J66, J67 (g)	Rats R32, R33 (g)	Jerboas J42, J51 (g)	Rats R31, R22 (g)
Per day:				
Total dry residue per animal	0·19	0·29	0·10	0·19
Fresh urine per animal	4·47	8·51	1·01	2·11
Water loss per animal	4·28	8·22	0·91	1·92
Fresh urine*	2·94	5·12	0·69	1·02
Dry residue*	0·12	0·17	0·07	0·09
Water loss*	2·82	4·95	0·62	0·93
Percentage				
of fresh weight	95·9	96·68	89·7	91·1
of dry weight	2337	2915	874·6	1026·3

* per 100 g body weight per day

APPENDIX 9

BODY TEMPERATURE OF THE JERBOA AND WHITE RAT

White rat (Wistar)

Date: March 28, 1954, at 1400 h
Environmental temperature: 19·5° C
Relative humidity: 68 per cent

	Series	Sex	Weight (g)	Body temperature (°C)
	1	Male	167	37·5
	2		213	37·9
	3		225	37·4
	4		190	37·8
	5		196	36·8
	6		230	37·8
Mean				37·5
	7	Female	215	37·8
	8		170	38·1
	9		143	37·0
	10		125	37·5
	11		135	37·7
	12		220	37·8
Mean				37·6
General mean				**37·55**

Jerboas on wet diet

Date (1954)	Ext. temp. (°C)	Rel. hum. (%)	Body temperature (°C)				Mean temp. (°C)
Male jerboas			J29	J32	J30	J33	
21/2	18	70	37·0	37·0	—	—	
07/3	10	75	—	37·0	—	36·6	
28/3	20	63	36·8	37·1	—	—	
04/4	20	68	36·8	36·8	36·9	36·8	
11/4	20·5	68	36·8	—	37·2	37·4	
09/5	22	68	36·8	36·7	37·0	36·9	
23/5	23	66	36·5	37·4	—	—	
07/6	25	65	36·5	—	37·1	36·4	
25/6	27	73	36·4	—	37·0	—	
25/9	26	60	36·8	37·0	36·8	36·8	
Mean			36·71	37·0	37·0	36·81	36·9

BODY TEMPERATURE OF THE JERBOA AND WHITE RAT

Date (1954)	Ext. temp. (°C)	Rel. hum. (%)	Body temperature (°C)				Mean temp. (°C)
Female jerboas			J20	J28	J36	J21	
21/2			—	—	37·5	36·8	
07/3			37·5	—	—	37·4	
28/3			37·6	36·3	—	—	
04/4			37·0	36·7	—	37·1	
11/4			—	36·7	—	37·3	
09/5			—	36·6	37·4	37·3	
23/5			—	36·8	37·2	—	
07/6			—	37·4	—	37·4	
25/6			37·0	36·8	37·4	36·9	
25/9			37·6	36·8	36·8	37·2	
Mean			37·34	36·76	37·26	37·13	37·1

Jerboas on dry diet

Male jerboas			J38	J25	J19	J27	
21/2			—	—	—	36·0	
07/3			37·0	36·7	—	—	
28/3			—	—	—	37·2	
04/4			36·6	—	37·0	36·5	
11/4			36·8	36·2	36·8	36·5	
09/5			36·8	—	36·8	36·5	
23/5			—	—	36·9	—	
07/6			36·3	36·9	—	—	
25/6			36·8	36·9	—	36·7	
25/9			36·4	36·7	36·5	37·1	
Mean			36·68	36·68	36·80	36·64	36·7
Female jerboas			J15	J35	J3	J18	
21/2			—	36·8	37·0	36·8	
07/3			—	—	37·7	—	
28/3			37·6	—	36·5	36·6	
04/4			—	—	37·5	36·5	
11/4			—	36·9	—	—	
09/5			—	—	—	—	
23/5			36·9	—	—	36·5	
07/6			36·6	—	—	36·6	
25/6			36·9	37·2	—	36·6	
25/9			36·4	36·4	—	36·7	
Mean			36·88	36·82	37·17	36·61	36·9

Summary: Mean of 8 jerboas on wet diet: weight 151 g, body temperature 37° C
Mean of 8 jerboas on dry diet: weight 155 g, body temperature 36·8° C

Appendix 10

ZONE OF BODY-TEMPERATURE REGULATION

Jerboas on wet diet

Mean weight 154 g

Date (1955)	Ext. temp. (°C)	Rel. hum. (%)	Atm. press. (mm Hg)	Exp. temp. (°C)	Body temperature (°C)			Mean temp. (°C)
					J29 Male	J45 Female	J43 Male	
26/1	17	67	765·6	−2	37·72	36·06	36·56	
29/1	18	62	762·5	−2	35·83	37·61	36·11	36·65
30/1	18·5	68	763·5	0	37·72	37·83	37·17	
31/1	18·5	66	765·3	0	37·28	37·11	38·06	37·53
01/2	17·5	66	765·8	5	36·67	36·28	37·06	
02/2	16·5	67	764·0	5	37·78	36·67	37·22	36·95
03/2	16·0	72	760·9	10	37·33	36·39	37·28	
04/2	17·0	66	760·4	10	36·39	—	37·39	36·95
07/2	16·5	64	765·8	15	37·28	36·35	37·56	
08/2	16·5	72	757·4	15	—	36·22	37·28	36·95
10/2	17·0	66	760·7	20	37·17	36·53	36·72	
11/2	16·5	67	763·2	20	36·89	37·44	36·89	36·95
12/2	17·0	60	764·8	25	36·50	36·78	36·89	
13/2	17·5	65	761·4	25	36·94	36·50	36·94	36·76
14/2	17·5	68	762·5	30	36·94	37·22	36·22	
15/2	17·5	60	762·0	30	36·11	36·11	36·50	36·35
16/2	17·5	63	765·0	35	37·83	37·28	37·50	
17/2	17·5	65	766·5	35	37·56	37·17	37·50	37·48
Mean								36·95

Jerboas on dry diet

Mean weight 155 g

Date	Ext. temp. (°C)	Rel. hum. (%)	Atm. press. (mm Hg)	Exp. temp. (°C)	Body temperature (°C)		Mean temp. (°C)
					J22 Female	J46 Male	
26/1				−2	35·56	36·94	
29/1				−2	36·89	38·44	36·96
30/1				0	36·89	37·67	
30/1				0	37·17	37·28	37·25
01/2				5	36·39	36·44	
02/2				5	36·67	37·00	36·62
03/2				10	37·06	36·28	
04/2				10	—	—	36·67
07/2				15	37·11	36·50	
08/2				15	36·89	35·67	36·54
10/2				20	36·89	36·28	
11/2				20	—	—	36·58

ZONE OF BODY-TEMPERATURE REGULATION

Date	Ext. temp. (°C)	Rel. hum. (%)	Atm. press. (mm Hg)	Exp. temp. (°C)	Body temperature (°C)		Mean temp. (°C)
					J22 Female	J46 Male	
12/2				25	37·61	—	
13/2				25	37·72	37·00	37·44
14/2				30	37·39	37·89	
15/2				30	37·61	37·17	37·51
16/2				35	37·61	38·11	
17/2				35	37·78	38·50	38·00
Mean							37·06

Summary: For environmental temperatures from −2 to 35° C
    mean for 3 jerboas on wet diet: weight 154 g, body temperature 36·95° C,
    mean for 2 jerboas on dry diet: weight 155 g, body temperature 37·06° C.

Zone of body-temperature regulation of jerboas and white rats

Jerboas

Mean weight 149 g

Date (1955)	Ext. temp. (°C)	Rel. hum. (%)	Atm. press. (mm Hg)	Exp. temp. (°C)	Body temperature (°C)			Mean temp. (°C)
					J50 Fem.	J42 Male	J49 Male	
16/6	27·5	57	760·2	−2	37·28	37·72	38·56	
18/6	26·5	60	759·9	−2	38·39	38·17	37·11	37·87
19/6	26·5	63	759·2	0	37·83	38·00	37·67	
20/6	26·5	65	759·2	0	38·33	38·28	38·00	38·01
21/6	27·0	70	759·4	5	37·22	37·22	37·44	
22/6	27·0	68	758·9	5	—	37·33	37·67	37·38
23/6	27·0	67	756·6	10	36·89	36·94	37·72	
24/6	27·0	70	756·9	10	37·06	37·44	38·33	37·39
25/6	26·5	67	758·7	15	36·89	36·89	37·72	
27/6	26·5	63	758·9	15	37·61	37·06	37·67	37·30
28/6	26·5	66	758·4	20	37·72	36·56	37·22	
29/6	27·0	69	757·4	20	37·61	37·06	37·72	37·31
01/7	26·6	62	757·4	25	37·44	37·11	37·33	
02/7	26·5	65	758·1	25	—	—	—	37·29
04/7	27·5	68	757·4	30	37·67	37·78	37·06	
05/7	27·0	68	756·1	30	37·28	37·56	36·50	
24/9	26·0	65	762·0	30	37·50	37·39	37·33	
25/9	26·0	75	763·2	30	37·06	36·89	36·89	37·24
06/7	27·5	68	756·4	35	37·61	37·39	37·11	
08/7	27·5	67	758·1	35	37·83	37·28	37·78	
26/9	26·5	65	762·0	35	37·61	37·89	38·06	
27/9	26·0	60	759·9	35	37·61	37·89	38·00	37·67
28/9	26·0	58	761·7	40	39·50	39·67	39·56	
01/10	26·0	70	760·4	40	38·00	39·89	39·89	39·42
Mean								37·69

APPENDIX 10

White rats

Mean weight 172 g

Date (1955)	Ext. temp. (°C)	Rel. hum. (%)	Atm. press. (mm Hg)	Exp. temp. (°C)	Body temperature (°C)			Mean temp. (°C)
					R1 *Male*	R2 *Male*	R3 *Male*	
16/6	27·5	57	760·2	−2	37·44	37·28	37·44	
18/6	26·5	60	759·9	−2	37·22	38·06	38·06	37·58
19/6	26·5	63	759·2	0	36·94	38·00	38·00	
20/6	26·5	65	759·2	0	37·22	38·00	37·89	37·67
21/6	27·0	70	759·4	5	36·89	38·00	38·28	
22/6	27·0	68	758·9	5	36·72	37·89	38·22	37·66
23/6	27·0	67	756·6	10	37·56	37·50	38·28	
24/6	27·0	70	756·9	10	37·22	—	—	37·64
25/6	26·5	67	758·7	15	37·50	38·11	37·72	
27/6	26·5	63	758·9	15	37·11	37·67	37·61	37·62
28/6	26·5	66	758·4	20	37·17	37·50	38·11	
29/6	27·0	69	757·4	20	37·44	37·67	37·89	37·63
01/7	26·6	62	757·4	25	—	37·67	37·72	
02/7	26·5	65	785·1	25	37·06	37·61	37·56	37·52
04/7	27·5	68	757·4	30	37·56	37·61	37·72	
05/7	27·0	68	756·1	30	37·11	37·33	37·56	
24/9	26·0	65	762·0	30	—	—	—	
25/9	26·0	73	763·2	30	—	—	—	37·84
06/7	27·5	68	756·4	35	38·78	38·50	38·33	
08/7	27·5	67	758·1	35	39·06	38·61	38·94	
26/9	26·5	65	762·0	35	38·72	38·72	38·56	
27/9	26·0	60	759·9	35	38·50	38·72	38·56	38·66
28/9	26·0	58	761·7	40	40·44	42·06	40·28	
01/10	26·0	70	760·4	40	41·22	42·06	40·78	41·14
Mean								**38·06**

Appendix 11

ENERGY METABOLISM IN RELATION TO EXTERNAL TEMPERATURE

Respiratory exchanges of jerboas on wet diet

Date	Animal	Weight (g)	Exper. temp. (°C)	O_2 (ml/g/h)	R.Q.	kcal/kg/h
18.11.55	J32M	142	0	4·18	0·79	20·0
	J33M	138	0	3·66	0·77	17·4
	J38M	155	0	3·44	0·75	16·3
25.08.56	J42M	145	0	3·85	0·74	18·2
28.04.57	J47M	147	0	3·28	0·76	15·6
Mean				**3·68**		**17·5**
16.11.55	J33M	144	5	3·52	0·78	16·8
	J32M	144	5	3·29	0·76	15·6
01.09.56	J42M	155	5	2·81	0·78	13·4
04.05.57	J59F	159	5	3·13	0·84	14·2
02.05.57	J47M	147	5	2·67	0·82	12·9
Mean				**3·09**		**14·8**
14.11.55	J32M	147	10	2·55	0·74	12·0
	J33M	148	10	2·56	0·79	12·3
	J38M	157	10	2·35	0·73	11·1
25.05.56	J42M	142	10	2·49	0·79	11·9
08.05.57	J59F	156	10	2·41	0·82	11·6
21.04.58	J47M	153	10	2·40	0·72	11·3
Mean				**2·46**		**11·7**
12.11.55	J33M	145	15	2·29	0·74	10·8
	J38M	161	15	2·18	0·75	10·3
	J32M	160	15	1·87	0·77	8·94
23.05.56	J42M	143	15	2·09	0·73	9·85
28.09.57	J47M	163	15	1·89	0·78	9·07
Mean				**2·06**		**9·82**
09.11.55	J32M	150	20	1·55	0·71	7·28
	J33M	150	20	1·63	0·72	7·70
	J38M	166	20	1·50	0·73	7·12
08.04.56	J42M	128	20	2·09	0·75	7·43
21.05.57	J59F	157	20	1·62	0·79	7·77
12.04.58	J47M	157	20	1·56	0·88	7·70
Mean				**1·66**		**7·50**
07.11.55	J38M	170	25	1·28	0·71	6·01
	J33M	155	25	1·20	0·79	5·79
	J32M	155	25	1·26	0·76	5·99
14.04.56	J42M	153	25	1·18	0·76	5·64
27.10.57	J47M	162	25	1·18	0·79	5·66
Mean				**1·22**		**5·82**

APPENDIX II

Date	Animal	Weight (g)	Exper. temp. (°C)	O$_2$ (ml/g/h)	R.Q.	kcal/kg/h
06.08.56	J42M	148	28	0·86	0·89	4·23
12.04.56	J42M	131	28	1·22	0·74	5·78
11.07.56	J42M	137	28	0·98	0·92	4·89
Mean				**1·02**		**4·96**
17.05.56	J42M	144	29	1·00	0·75	4·77
05.07.56	J42M	138	29	0·86	0·92	4·30
04.08.56	J42M	148	29	0·79	0·88	3·90
Mean				**0·89**		**4·32**
01.11.55	J38M	170	30	0·79	0·77	3·80
	J32M	155	30	0·80	0·74	3·82
	J33M	168	30	0·93	0·73	4·41
	J28F	126	30	0·94	0·81	4·55
	J59F	184	30	0·77	0·81	3·74
	J42M	142	30	0·83	0·75	3·95
	J47M	166	30	0·74	0·79	3·57
Mean				**0·83**		**3·98**
31.07.56	J42M	147	31	0·82	0·75	3·90
02.01.57	J47M	153	31	0·95	0·72	4·49
Mean				**0·88**		**4·19**
08.11.56	J33M	141	32	0·85	0·73	4·02
06.05.58	J47M	151	32	0·96	0·75	4·57
Mean				**0·90**		**4·30**
25.10.55	J38M	175	33	0·86	0·74	4·07
	J33M	168	33	0·91	0·72	4·30
26.10.55	J59F	182	33	0·84	0·75	4·02
	J36F	150	33	0·85	0·76	4·08
	J28F	120	33	0·96	0·80	4·73
01.11.56	J42M	175	33	0·92	0·77	4·39
13.10.55	J21F	118	33	0·85	0·83	4·16
10.01.57	J47M	149	33	1·06	0·72	5·03
	J49M	141	33	1·03	0·75	5·18
Mean				**0·92**		**4·43**
22.06.56	J42M	147	34	0·91	0·79	4·41
19.10.55	J49M	140	35	1·03	0·80	4·94
	J42M	153	35	0·94	0·91	4·67
21.10.55	J38M	184	35	0·89	0·80	4·31
	J32M	162	35	0·98	0·87	4·82
	J33M	177	35	0·92	0·87	4·52
22.10.55	J36F	155	35	0·82	0·75	4·05
	J59F	185	35	0·89	0·80	4·31
	J28F	125	35	1·01	0·81	4·86
Mean				**0·93**		**4·56**
17.10.58	J47M	155	40	1·01	0·81	5·32
07.01.59	J70M	184	40	1·23	0·75	5·86
Mean				**1·12**		**5·59**
21.10.58	J47M	152	45	1·47	0·70	6·91

ENERGY METABOLISM IN RELATION TO EXTERNAL TEMPERATURE

Respiratory exchanges of jerboas on dry diet

Date	Animal	Weight (g)	Exper. temp. (°C)	O_2 (ml/g/h)	R.Q.	kcal/kg/h
25.08.56	J62M	145	0	3·85	0·72	18·13
30.04.57	J32M	152	0	3·27	0·90	16·11
27.04.58	J51M	132	0	3·26	0·76	15·53
Mean				**3·46**		**16·59**
27.05.56	J62M	142	5	3·38	0·80	16·25
04.05.57	J32M	149	5	2·72	0·75	12·95
07.10.57	J51M	136	5	2·59	0·75	12·29
Mean				**2·90**		**13·83**
25.05.56	J62M	142	10	2·54	0·73	11·99
08.05.57	J32M	145	10	2·35	0·81	11·35
21.04.58	J51M	135	10	2·37	0·70	11·13
Mean				**2·42**		**11·49**
18.08.56	J62M	149	15	2·01	0·77	9·60
24.08.58	J51M	149	15	2·02	0·72	9·51
Mean				**2·01**		**9·55**
16.08.56	J62M	149	20	1·50	0·75	7·12
17.05.58	J51M	143	20	1·62	0·82	7·83
Mean				**1·56**		**7·47**
14.08.56	J62M	151	25	1·14	0·79	5·47
26.05.57	J32M	140	25	1·17	0·78	5·63
30.04.58	J51M	133	25	1·30	0·77	6·24
Mean				**1·20**		**5·78**
11.07.56	J62M	150	28	0·88	0·84	4·31
06.02.56	J62M	151	28	0·85	0·70	3·99
Mean				**0·87**		**4·15**
17.05.56	J62M	145	29	0·89	0·78	4·26
05.07.56	J62M	154	29	0·80	0·78	3·85
Mean				**0·85**		**4·06**
03.07.56	J62M	155	30	0·69	0·88	3·43
02.08.56	J62M	156	30	0·66	0·85	3·22
Mean				**0·68**		**3·32**
22.05.58	J51M	145	30	0·86	0·74	4·07
02.10.58	J51M	144	30	0·82	0·71	3·86
Mean				**0·84**		**3·97**
15.05.56	J62M	145	31	0·84	0·85	4·10
21.05.56	J62M	144	31	0·74	0·88	3·63
01.07.56	J62M	157	31	0·72	0·81	3·50
Mean				**0·77**		**3·74**
10.05.58	J51M	135	31	0·88	0·78	4·21
16.05.56	J62M	133	32	1·03	0·80	5·00
29.06.56	J62M	160	32	0·71	0·86	3·48
08.01.57	J51M	139	32	0·97	0·82	4·70
08.09.57	J51M	146	32	0·77	0·89	3·80
25.03.57	J32M	167	32	0·95	0·77	4·56
Mean				**0·90**		**4·35**

APPENDIX II

Date	Animal	Weight (g)	Exper. temp. (°C)	O_2 (ml/g/h)	R.Q.	kcal/kg/h
13.05.56	J62M	147	33	1·04	0·80	5·02
24.06.56	J62M	153	33	0·86	0·71	4·05
12.01.57	J51M	138	33	1·11	0·75	5·29
17.06.57	J51M	152	33	1·08	0·71	5·10
Mean				**1·02**		**4·86**
18.04.56	J62M	133	34	1·06	0·81	5·14
11.05.56	J62M	149	35	1·25	0·75	5·95
14.10.58	J51M	136	35	0·99	0·75	4·73
Mean				**1·12**		**5·34**
17.10.58	J51M	136	40	1·14	0·78	5·48
07.01.59	J51M	144	40	1·30	0·75	6·18
Mean				**1·22**		**5·83**
21.10.58	J51M	134	45	1·53	0·70	7·18

Respiratory exchanges of white rats

Date	Animal	Weight (g)	Exper. temp. (°C)	O_2 (ml/g/h)	R.Q.	kcal/kg/h
17.11.55	R5M	148	0	4·04	0·79	19·38
	R1F	144	0	3·67	0·85	17·88
	R3F	154	0	3·57	0·74	21·62
25.08.56	R8F	150	0	3·17	0·83	15·38
27.04.58	R29F	157	0	3·72	0·77	17·44
Mean				**3·83**		**18·34**
15.11.55	R5M	149	5	3·95	0·73	18·68
	R1F	144	5	3·00	0·82	14·50
	R3F	154	5	3·28	0·74	15·54
02.09.56	R9M	162	5	2·81	0·88	13·78
02.05.57	R16F	162	5	3·81	0·73	18·01
24.04.58	R29F	159	5	3·24	0·70	15·20
Mean				**3·35**		**15·95**
12.11.55	R2M	160	10	2·32	0·82	11·19
	R1F	142	10	2·64	0·76	12·58
	R3F	153	10	2·88	0·75	13·69
25.05.56	R7M	155	10	2·44	0·76	11·64
21.08.56	R8F	150	10	2·69	0·76	12·81
21.04.58	R29F	161	10	2·62	0·75	12·48
Mean				**2·60**		**12·40**
10.11.55	R2M	158	15	1·72	0·73	8·13
	R1F	153	15	2·18	0·76	10·36
	R3F	142	15	2·12	0·77	10·11
02.05.56	R7M	157	15	1·93	0·81	9·32
18.09.56	R8F	152	15	2·13	0·83	10·36
24.08.58	R29F	165	15	2·35	0·76	11·19
Mean				**2·07**		**9·91**

ENERGY METABOLISM IN RELATION TO EXTERNAL TEMPERATURE

Date	Animal	Weight (g)	Exper. temp. (°C)	O_2 (ml/g/h)	R.Q.	kcal/kg/h
08.11.55	R2M	160	20	1·53	0·77	7·30
	R1F	142	20	1·87	0·72	8·85
	R3F	154	20	1·89	0·76	8·52
06.04.56	R7M	172	20	1·76	0·71	8·30
16.08.56	R8F	154	20	1·64	0·75	7·80
19.03.57	R16F	168	20	1·74	0·76	8·29
15.04.58	R29F	163	20	1·47	0·79	7·05
Mean				**1·69**		**8·02**
06.10.55	R2M	180	25	1·57	0·74	7·47
05.11.55	R3F	153	25	1·49	0·76	7·10
	R1F	144	25	1·31	0·82	6·36
06.04.56	R7M	178	25	1·38	0·74	6·56
26.05.57	R16F	176	25	1·29	0·79	6·12
30.04.58	R29F	164	25	1·47	0·75	6·99
Mean				**1·42**		**6·77**
06.08.56	R8F	155	28	1·39	0·77	6·65
11.07.56	R8F	158	28	1·38	0·83	6·71
Mean				**1·39**		**6·68**
05.07.56	R8F	167	29	1·31	0·74	6·21
04.08.56	R8F	160	29	1·46	0·73	6·91
Mean				**1·38**		**6·56**
03.11.55	R2M	160	30	1·41	0·79	6·80
	R1F	147	30	1·13	0·79	5·46
	R3F	156	30	1·20	0·86	5·86
14.04.56	R7M	175	30	1·33	0·76	6·34
02.08.56	R8F	162	30	1·35	0·78	6·47
30.10.57	R19F	219	30	1·15	0·80	5·54
30.12.56	R10M	151	30	1·31	0·72	6·17
	R13M	207	30	1·36	0·72	6·43
23.03.57	R16F	166	30	1·29	0·73	6·10
03.05.58	R29F	165	30	1·34	0·74	6·35
Mean				**1·29**		**6·15**
31.07.56	R8F	167	31	1·35	0·74	6·39
22.09.56	R6F	141	31	1·51	0·73	7·17
04.01.57	R10M	149	31	1·39	0·76	6·63
	R13M	204	31	1·42	0·73	6·73
10.05.58	R29F	159	31	1·29	0·79	6·19
Mean				**1·39**		**6·62**
16.04.56	R7M	167	32	1·58	0·77	7·55
29.07.56	R8F	171	32	1·32	0·77	7·06
08.09.56	R10M	120	32	1·31	0·79	6·32
20.09.56	R6F	146	32	1·39	0·70	6·53
08.01.57	R13M	194	32	1·48	0·71	6·99
27.03.57	R16F	168	32	1·26	0·72	5·95
06.05.58	R29F	160	32	1·51	0·75	7·17
Mean				**1·39**		**6·79**

APPENDIX II

Date	Animal	Weight (g)	Exper. temp. (°C)	O_2 (ml/g/h)	R.Q.	kcal/kg/h
15.10.55	R2M	175	33	1·44	0·73	6·82
27.10.55	R1F	148	33	1·50	0·71	7·05
	R3F	157	33	1·42	0·71	6·69
12.01.57	R10M	148	33	1·44	0·74	6·85
	R13M	197	33	1·50	0·75	7·14
06.09.57	R9M	158	33	1·40	0·73	6·62
18.09.57	R6F	155	33	1·38	0·74	6·53
27.07.56	R8F	174	33	1·31	0·74	6·22
24.06.56	R7M	137	33	1·43	0·83	6·93
16.09.57	R19F	190	33	1·72	0·70	8·07
Mean				**1·45**		**6·89**
18.04.56	R7M	164	34	1·85	0·77	8·85
22.06.56	R7M	142	34	1·46	0·84	7·10
Mean				**1·65**		**7·98**
20.10.55	R2M	182	35	1·67	0·86	8·15
	R1F	152	35	1·45	0·88	7·15
	R3F	170	35	1·40	0·87	6·88
11.05.56	R7M	171	35	1·91	0·80	10·28
Mean				**1·61**		**8·12**
24.04.56	R7M	147	40	2·41	1·00	12·20
17.10.58	R29F	155	40	2·40	0·70	9·58
Mean				**2·40**		**10·90**

APPENDIX 12

WATER LOSS OF JERBOAS AND RATS

Insensible perspiration and evaporative water loss of jerboas on wet diet

Date	Animal	Weight (g)	Exper. temp. (°C)	O$_2$/kg/h (g)	R.Q.	kcal/kg/h	H$_2$O/kg/h (g)
28.04.57	J47M	147	0	4·69	0·76	15·64	1·14
02.04.58	J47M	147	0	4·55	0·71	14·93	1·25
Mean			0	4·62		15·29	1·20
07.10.57	J47M	*153*	5	*3·47*	*0·73*	*11·45*	*1·04*
06.05.57	J47M	149	10	3·15	0·72	10·38	1·08
21.08.58	J47M	165	10	3·41	0·71	11·22	0·86
Mean			10	3·28		10·80	0·97
10.05.57	J47M	149	15	2·96	0·75	9·84	1·01
12.05.57	J59F	155	15	3·08	0·74	10·22	0·80
Mean			15	3·02		10·03	0·90
23.09.57	J47M	161	20	2·11	0·76	7·05	0·70
21.05.57	J59F	157	20	2·31	0·79	7·77	0·79
Mean			20	2·21		7·41	0·72
27.10.57	J47M	162	25	1·69	0·79	5·66	0·73
30.04.58	J47M	149	25	1·77	0·81	5·99	0·68
Mean			25	1·73		5·83	0·71
25.08.57	J47M	165	30	1·24	0·73	4·10	0·70
20.08.57	J47M	166	30	1·06	0·79	3·57	0·56
22.05.58	J47M	160	30	1·41	0·77	4·71	0·87
Mean			30	1·24		4·13	0·71
03.09.57	J47M	168	31	1·22	0·72	4·03	0·86
10.05.58	J47M	152	31	1·32	0·76	4·41	1·34
Mean			31	1·27		4·22	1·10
02.01.57	J64M	137	31	1·65	0·77	5·53	1·22
02.01.57	J49M	150	31	1·68	0·72	5·55	0·93
Mean			31	1·53		5·54	1·09
06.05.58	J47M	151	32	1·37	0·75	4·57	1·33
06.01.57	J49M	141	32	1·74	0·69	5·72	1·44
23.03.57	J49M	173	32	1·84	0·69	6·04	1·16
Mean			32	1·65		5·51	1·31
10.01.57	J47M	149	33	1·52	0·72	5·03	1·55
10.01.57	J49M	141	33	1·56	0·75	5·18	1·23
Mean			33	1·54		5·10	1·39
14.10.58	J47M	*158*	*35*	*1·55*	*0·73*	*5·14*	*1·50*
17.10.58	J47M	155	40	1·57	0·81	5·32	8·07
07.01.59	J70M	184	40	1·76	0·75	5·86	8·46
Mean			40	1·67		5·59	8·26
21.10.58	J47M	152	45	2·11	0·75	6·91	14·21

APPENDIX 12

Insensible perspiration and evaporative water loss of jerboas on dry diet

Date	Animal	Weight (g)	Exper. temp. (°C)	O_2/kg/h (g)	R.Q.	kcal/kg/h	H_2O/kg/h (g)
04.10.57	J51M	133	0	4·49	0·70	14·76	0·92
24.04.58	J51M	132	0	4·66	0·76	15·53	0·96
Mean			0	4·58		15·14	0·94
04.05.57	J32M	149	5	3·90	0·75	12·95	0·69
07.10.57	J51M	136	5	3·70	0·75	12·29	0·99
Mean			5	3·80		12·62	0·82
02.08.58	J51M	*152*	10	*3·72*	*0·70*	*12·22*	*0·81*
18.04.57	J51M	135	15	3·07	0·70	10·07	0·93
24.08.58	J51M	149	15	2·89	0·72	9·51	0·62
Mean			15	2·98		9·79	0·78
15.04.58	J51M	*137*	20	*2·45*	*0·76*	*8·19*	*0·68*
26.05.57	J32M	140	25	1·68	0·78	5·63	0·61
30.04.58	J51M	133	25	1·87	0·77	6·24	0·65
Mean			25	1·77		5·94	0·63
22.05.58	J51M	145	30	1·23	0·74	4·07	0·59
25.08.57	J51M	149	30	1·45	0·69	4·80	0·72
Mean			30	1·34		4·44	0·65
04.01.57	J51M	143	31	1·44	0·73	4·76	0·89
03.09.57	J51M	149	31	1·22	0·84	4·15	0·53
Mean			31	1·33		4·46	0·71
08.01.57	J51M	139	32	1·39	0·82	4·70	0·70
08.09.57	J51M	146	32	1.10	0·89	3·80	0·84
23.08.57	J32M	167	32	1·36	0·77	4·56	0·77
Mean			32	1·30		4·41	0·77
12.01.57	J51M	138	33	1·59	0·75	5·29	1·23
17.06.57	J51M	152	33	1·55	0·71	5·10	0·92
Mean			33	1·57		5·19	1·07
14.10.58	J51M	*136*	35	*1·42*	*0·75*	*4·73*	*1·37*
17.10.58	J51M	*136*	40	*1·63*	*0·78*	*5·48*	*5·06*
21.10.58	J51M	134	45	2·19	0·70	7·18	13·79

Insensible perspiration and evaporative water loss of white rats

Date	Animal	Weight (g)	Exper. temp. (°C)	O_2/kg/h (g)	R.Q.	kcal/kg/h	H_2O/kg/h (g)
27.04.58	R29F	*157*	0	*5·32*	0·77	17·74	2·74
02.05.57	R16F	162	5	5·45	0·73	18·01	2·53
24.04.58	R29F	159	5	4·63	0·70	15·20	2·65
Mean			5	5·04		16·61	2·59

Date	Animal	Weight (g)	Exper. temp. (°C)	O_2/kg/h (g)	R.Q.	kcal/kg/h	H_2O/kg/h (g)
21.08.58	R29F	168	10	4·10	0·72	13·55	2·68
30.09.57	R19F	195	10	3·91	0·68	12·84	1·71
Mean			10	4·01		13·19	2·19
24.08.58	R29F	165	15	3·66	0·76	11·19	1·96
18.05.58	R29F	160	20	2·71	0·80	9·14	1·54
19.03.57	R16F	168	20	2·48	0·76	8·29	1·38
23.09.57	R19F	194	20	1·83	0·80	6·18	1·43
Mean			20	2·34		7·87	1·45
26.05.57	R16F	176	25	1·85	0·79	6·21	1·09
27.10.57	R19F	224	25	1·45	0·81	4·89	0·95
30.04.58	R29F	164	25	2·10	0·75	6·99	1·27
Mean			25	1·80		6·03	1·10
16.01.57	R10M	137	30	1·83	0·80	6·16	1·17
30.10.57	R19F	219	30	1·65	0·80	5·54	1·09
Mean			30	1·74		5·85	1·13
10.05.58	R29F	159	31	1·84	0·79	6·19	1·74
08.01.57	R10M	144	32	2·01	0·81	6·79	2·78
06.05.58	R29F	160	32	2·15	0·75	7·17	2·00
27.03.57	R16F	168	32	1·80	0·72	5·95	1·63
Mean			32	1·99		6·64	2·13
17.06.57	R16F	176	33	1·65	0·84	5·61	3·20
16.09.57	R19F	190	33	2·46	0·70	8·07	2·01
Mean			33	2·05		6·84	2·60
14.10.58	R29F	150	35	2·33	0·74	7·73	3·03
17.10.58	R29F	155	40	2·92	0·70	9·58	14·74

Latent heat of evaporation in jerboas and rats in relation to total heat

Table based on the three preceding tables
(1 g H_2O = 0·59 kcal)

Jerboas on wet diet

Exp. temp. (°C)	Evap. H_2O (g/kg/h)	kalories equiv. (× 100)	Total heat (kcal/kg/h)	Ratio (%)
0	1·201	70·865	15·293	4·6
5	1·047	61·820	11·458	5·3
10	0·978	57·692	10·806	5·3
15	0·909	53·672	10·036	5·3
20	0·722	42·604	7·413	5·7
25	0·711	41·972	5·831	7·2
30	0·714	42·132	4·132	10·2

APPENDIX 12

Exp. temp. (°C)	Evap. H$_2$O (g/kg/h)	kcalories equiv. (× 100)	Total heat (kcal/kg/h)	Ratio (%)
31	1·091	64·369	5·103	12·2
32	1·316	77·679	5·516	14·1
33	1·395	82·317	5·108	16·1
35	1·502	88·659	5·146	17·2
40	8·268	487·812	5·590	87·2
45	14·217	838·844	6·913	121·3

Jerboas on dry diet

Exp. temp. (°C)	Evap. H$_2$O (g/kg/h)	kcalories equiv. (× 100)	Total heat (kcal/kg/h)	Ratio (%)
0	0·947	55·896	15·146	3·7
5	0·828	48·864	12·624	3·9
10	0·814	48·026	12·228	3·9
15	0·781	46·091	9·791	4·7
20	0·687	40·521	8·181	4·9
25	0·632	37·294	5·941	6·3
30	0·659	38·893	4·441	8·7
31	0·713	42·073	4·461	9·4
32	0·776	45·825	4·411	10·4
33	1·079	63·696	5·199	12·3
35	1·371	80·924	4·734	17·1
40	5·061	298·634	5·487	54·4
45	13·798	814·087	7·185	113·3

White rats on wet diet

Exp. temp. (°C)	Evap. H$_2$O (g/kg/h)	kcalories equiv. (× 100)	Total heat (kcal/kg/h)	Ratio (%)
0	2·749	162·232	17·744	9·1
5	2·594	153·028	16·611	9·2
10	2·199	129·758	13·197	9·8
15	1·968	116·135	11·191	10·4
20	1·453	85·762	7·876	10·8
25	1·105	65·212	6·036	10·8
30	1·136	67·041	5·857	11·4
31	1·746	103·025	6·194	16·6
32	2·139	126·212	6·641	19·0
33	2·605	153·689	6·847	22·4
35	3·032	178·929	7·738	23·1
40	14·742	869·819	9·581	90·7
45	Lethal			

WATER LOSS OF JERBOAS AND RATS

Evaporative water loss in relation to oxygen consumed

Exp. temp. (°C)	Jerboas on wet diet			Jerboas on dry diet			White rats on wet diet		
	H_2O (g/kg/h)	O_2 (g/kg/h)	H_2O/O_2	H_2O (g/kg/h)	O_2 (g/kg/h)	H_2O/O_2	H_2O (g/kg/h)	O_2 (g/kg/h)	H_2O/O_2
0	1·201	4·626	0·259	0·947	4·581	0·207	2·749	5·323	0·516
5	1·047	3·474	0·301	0·828	3·804	0·217	2·594	5·042	0·514
10	0·978	3·283	0·298	0·814	3·729	0·218	2·199	4·013	0·548
15	0·909	3·025	0·300	0·781	2·981	0·262	1·968	3·664	0·537
20	0·722	2·217	0·325	0·687	2·456	0·279	1·454	3·348	0·434
25	0·711	1·734	0·410	0·632	1·778	0·355	1·105	1·802	0·613
30	0·714	1·240	0·576	0·659	1·344	0·490	1·136	1·741	0·652
35	1·502	1·557	0·965	1·371	1·426	0·962	3·032	2·336	1·298
40	8·268	1·673	4·942	5·061	1·638	3·089	14·742	2·922	5·045
45	14·217	2·114	*6·725	13·798	2·191	*6·297		Lethal	

* See text (p. 69) for explanation.

BIBLIOGRAPHY

ADOLPH, E. F. (1933), The metabolism and distribution of water in body and tissues, *Physiol. Rev.*, 13, No. 3: 336–371.
ADOLPH, E. F. (1943), *Physiological Regulations*, The Jaques Cattell Press, Lancaster, Penn.
ADOLPH, E. F. (1943), Physiological fitness for the desert, *Fed. Proc.*, 2, No. 3: 158–164.
ADOLPH, E. F. (1947), Tolerances to heat and dehydration in several species of mammals, *Amer. J. Physiol.*, 151/2: 564–575.
ADOLPH, E. F. (1949), Desert, in *Physiology of Heat Regulation*, W. B. Saunders Company, Philadelphia, p. 330.
ADOLPH, E. F. et al. (1947), *Physiology of Men in the Desert*, Interscience Publishers, Inc., New York, p. 357.
ALEZAIS, H. (1900), *Contribution à la myologie des rongeurs* (Th. Sc. Nat. 1900–1901, No. 1038, série Ar. No. 371), Paris.
ALLEE, W. C., PARK, O., EMERSON, A. E., PARK, T. and SCHMIDT, P. (1950), *Principles of Animal Ecology*, W. B. Saunders Company, Philadelphia and London, p. 837.
AMBARD, D. (1914), *Physiologie normale et pathologique des reins*, Paris.
ANDERSON, J. (1902), *Zoology of Egypt: Mammalia* (revised and completed by W. E. DE WINTON), Hugh Rees, Ltd., London.
AUFRÈRE, L. et al. (1938), *La vie dans la région désertique nord-tropicale de l'ancien monde*, Paul Lechevalier, Paris, p. 405 (Société de Biogéographie VI).
BABINAU, L. and PAGÉ, E. (1955), On body fat and body water in rats, *Canad. J. Biochem. Physiol.*, 33: 970.
BAGNOLD, R. A. (1954), The physical aspects of dry deserts, in *Biology of Deserts*, Institute of Biology, London, pp. 7–12.
BAKER, D. G. and SELLERS, E. A. (1953), Carbohydrate metabolism in the rat exposed to a low environmental temperature, *Amer. J. Physiol.*, 174: 459–461.
BAKER, D. G. and SELLERS, E. A. (1957), Electrolyte metabolism in the rat exposed to a low environmental temperature, *Canad. J. Biochem. Physiol.*, 35: 631.
BALL, J. (1939), *Contributions to the Geography of Egypt*, Government Press, Bulac, Cairo, p. 308.
BARBOUR, H. G. (1920), *Physiol. Rev.*, 1: 295–326.
BARBOUR, H. G., MCKAY, E. A. and GRIFFITH, W. P. (1943), *Amer. J. Physiol.*, 140: 9–19.
BARNETT, L. (1954), The land of the Sun, *Life*, 16, No. 9 (International Edition): 28–49.
BASS, D. E. and BENSCHEL, A. (1956), Responses of body-fluid compartments to heat and cold, *Physiol. Rev.*, 36: 128–144.
BASS, D. E., KLEEMAN, C. R., QUIN, M., BENSCHEL, A. and HEGNAUER, A. H. (1955), Mechanics of acclimatization to heat in man, *Medicine, Baltimore*, 34: 323–380.
BAZETT, H. C. (1927), Review of temperature regulations, *Physiol. Rev.*, 7: 531.
BAZETT, H. C. (1949), The regulation of body temperature, Chapter 4 in *Physiology of Heat Regulation* (edited by L. B. NEWBURGH), W. B. Saunders Company, Philadelphia and London, pp. 109–192.
BEDFORD, T. (1948), *Basic Principles of Ventilation and Heating*, H. K. Lewis and Co., London.
BELGRAVE, C. D. (1923), *Siwa—The oasis of Jupiter Ammon*, John Lane, the Bodley Head Ltd., London, p. 275.
BENEDICT, F. G. (1938), *Vital Energetics: A Study in Comparative Basal Metabolism*, Carnegie Institution, Washington, D.C., p. 215.

BIBLIOGRAPHY

BENEDICT, F. G. and MACLEOD, G. (1928–29), The heat production of the albino rat, *J. Nutr.*, 1: 343–366, 367–398.
BERNARD, C. (1876), *Leçons sur la chaleur animale*, Baillière, Paris.
BERNARD, C. (1878), *Leçons sur les phénomènes de la vie*, Baillière, Paris.
BETIER, J. (1958), L'énergie solaire ressource inépuisable du désert, *Sci. et vie*, June 1958, Paris, p. 136.
BIEL, E. R. (1944), *Climatology of the Mediterranean Area*, University of Chicago Press, Chicago, Ill., p. 180.
BLAGDEN, C. (1775), *Phil. Trans.*, 13: 604.
BLUM, H. F. (1945), *Physiol. Rev.*, 25: 483.
BODENHEIMER, F. S. (1953), Problems of animal ecology and physiology in deserts, *Proc. Int. Symp. Desert Res., Jerusalem*, pp. 205–229.
BODENHEIMER, F. S. (1954), Problems of physiology and ecology of desert animals, in *Biology of Deserts*, Institute of Biology, London, pp. 162–167.
BOHNENKAMP, H. and PASQUAY, W. (1931), *Pflügers Arch. ges. Physiol.*, 228–79.
BOURLIÈRE, F. (1951), *Vie et mœurs des mammifères*, Payot, Paris, p. 250.
BREHM, A. E. von (1890), Die Säugetiere, in *Brehms Tierleben*, Bibliographisches Institut, Leipzig and Vienna.
BRODY, S. (1945), *Bioenergetics and Growth*, Reinhold Publishing Corp., New York, p. 1023.
BROWMAN, L. G. (1952), Artificial sixteen-hour day activity rhythms in the white rat, *Amer. J. Physiol.*, 168, 3: 694.
BURCH, G. E. and SODEMAN, W. A. (1942), Evaporation in temperature regulation, *J. clin. Invest.*, 21: 638.
BURCKHARDT, E., DONTCHEFF, L. and KAYSER, Ch. (1933), *Ann. Physiol. Physicochim. biol.*, 9: 303.
BURTON, A. C. (1939), Temperature regulation, *Annu. Rev. Physiol.*, 1: 109–130.
BUXTON, P. A. (1955), *Animal Life in Deserts*, Edward Arnold Ltd., London, p. 176.
BYKOV, C. (1956), Thermorégulation, in *L'écorce cérébrale et les organes internes*, Editions en langues étrangères, Moscow, Chapter IX, pp. 216–248.
CAILLEUX, A. (1952), *La Géologie*, Presses Universitaires de France, Paris, p. 126.
CAILLEUX, A. (1953), *Biogéographie mondiale*, Presses Universitaires de France, Paris, p. 126.
CAPOT-REY, R. (1953), *Le Sahara Français*, Presses Universitaires de France, Paris, p. 564.
CARPENTER, T. M. (1948), *Tables, factors and formulas for computing respiratory exchange and biological transformations of energy*, Carnegie Institution, publication 303-C, Washington, D.C., p. 147.
CASTELLANI, A. (1938), *Climate and Acclimatization*, John Bale, Sons and Curnow Ltd., London.
CAUVET, Commandant (1925), *Le Chameau*, J. B. Baillière et Fils, Paris, p. 782.
CAUVET, Commandant (1926), *Le Chameau, histoire, réligion, littérature, art*, J. B. Baillière et Fils, Paris, p. 207.
CHEVILLARD, L. (1935), *Ann. Physiol. Physicochim. biol.*, 11: 461.
CHEW, R. M. (1961), Water metabolism of desert-inhabiting vertebrates, *Biol. Rev.*, 36, No. 1: 1–31.
CLOUDSLEY-THOMPSON, J. L. (1954), *Biology of Deserts*, The proceedings of a symposium on the biology of hot and cold deserts organized by the Institute of Biology, London, p. 224.
CUÉNOT, L. (1951), *L'évolution biologique*, Masson et Cie, Paris.
CURASSON, G. (1947), *Le chameau et ses maladies*, Vigot Frères, Paris.
DAVIS, J. E. and VAN DYKE, H. B. (1932), *J. biol. Chem.*, 95: 73.
DAVIS, J. E. and HASTINGS, A. B. (1934), The measurement of oxygen consumption in immature rats, *Amer. J. Physiol.*, 109: 683–687.
DE COSSON, A. (1935), *Mareotis*, Country Life Ltd., London, p. 219.

BIBLIOGRAPHY

DEDENHAM, F. (1954), The geography of deserts, in *Biology of Deserts*, Institute of Biology, London, pp. 1–12.
DEIGHTON, T. (1933), Review of physical factors concerned in temperature regulation, *Physiol. Rev.*, 13: 422.
DEKEYSER, P.-L. and DERIVOT, J. (1959), *La vie animale au Sahara*, Collection Armand Colin No. 332, Section Biologie, Paris.
DÉROBERT, L. (1939), Coup de chaleur, in *Les troubles de la thermorégulation*, Masson et Cie, Paris, p. 218.
Desert Research (1953), Proceedings of the International Symposium on Desert Research held in Jerusalem, May 7–14, 1952: special publication No. 2, Jerusalem, 1953.
DILL, D. B. (1938), Physiological effects of hot climates and great heights, in *Life, Heat and Altitude*, Harvard University Press, Cambridge, Mass., p. 211.
DONTCHEFF, L. and KAYSER, C. (1934), Le rythme saisonnier du métabolisme de base chez le pigeon en fonction de la température moyenne du milieu, *Ann. Physiol. Physicochim. biol.*, 10: 285.
DU BOIS, A. and VAN DEN BERGHE, L. (1948), *Diseases of the Warm Climates*, William Heinemann, London.
DU BOIS, E. F. (1948), *Fever and the Regulation of Body Temperature*, Charles C. Thomas, Springfield, Ill.
DUHOT, E. (1945), *Les climats et l'organisme humain*, Presses Universitaires de France, Paris, p. 127.
EDHOLM, O. G. (1954), Physiological effects of cold environments on man, in *Biology of Deserts*, Institute of Biology, London, pp. 207–212.
EIJKMAN, C. (1924), Some questions concerning the influence of tropical climate on Man, *Lancet*, 206: 887.
ELLERMAN, J. R. (1949), *The families and genera of living rodents*, British Museum (Natural History), Vol. III, Part I, p. 124.
EMBERGER, L. (1938), La définition phytogéographique du climat désertique, in AUFRÈRE, L. et al., *La vie dans la région désertique nord-tropicale de l'ancien monde*, Paul Lechevalier, Paris, p. 13 (Société de Biogéographie VI).
FEDERER, W. T. (1955), *Experimental design*, in *Theory and Application*, The Macmillan Company, New York.
FINBERT, E. J. (1938), *La vie du chameau—le vaisseau du désert*, Albin Michel, Paris, p. 254.
FISHER, R. A. (1951), *The Design of Experiments*, Oliver and Boyd, London.
FIZES, M. (1755), *Traité des fièvres* (translated from Latin), Chez Besaint et Saillant, Paris.
FORBES, W. H. (1949), Laboratory and field studies, general principles, in *Physiology of Heat Regulation and the Science of Clothing*, W. B. Saunders Company, Philadelphia, p. 320.
FRY, F. E. J. (1958), Temperature compensation, *Annu. Rev. Physiol.*, 20: 207–224.
FURON, R. (1950), *Géologie de l'Afrique*, Payot, Paris, p. 250.
FURON, R. (1951), *Manuel de Préhistoire Générale*, Payot, Paris, p. 535.
GALVAO, P. E. (1947), Heat production in relation to body weight and body surface. Inapplicability of the surface law on dogs in the tropical zone, *Amer. J. Physiol.*, 148: 478.
GARDINIER, J.-P. (1958), L'alpinisme au Hoggar, *Sci. et vie*, June 1958, Paris, p. 132.
GAUTIER, E. F. (1950), *Le Sahara*, Payot, Paris, p. 231.
GAUTIER, E. F. (1952), Le passé de l'Afrique du Nord, in *Les siècles obscurs*, Payot, Paris, p. 457.
GEIGER, R. (1950), *The Climate near the Ground*, Harvard University Press, Cambridge Mass., p. 482.
GELINEO, S. (1934), Influence du milieu thermique d'adaptation sur la thermogenèse des homéothermes, *Ann. Physiol. Physicochim. biol.*, 10: 1083.

BIBLIOGRAPHY

GIAJA, A. (1931), *Ann. Physiol. Physicochim. biol.*, 7: 13.
GIAJA, J. (1938), *Homéothermie et thermorégulation*, Hermann et Cie, Paris, Vols. I and II, pp. 70 and 76.
GIAJA, J. and GELINEO, S. (1933), *Arch. int. Physiol.*, 37: 20.
GIAJA, J. and GELINEO, S. (1934), Alimentation et résistance au froid, *C.R. Acad. Sci., Paris*, 198: 2227.
GLICKMAN, N., HICK, F. K., KEETON, R. W. and MONTGOMERY, M. M. (1941), *Amer. J. Physiol.*, 134: 165–176.
GRADWOHL, R. B. M. (1948), *Clinical Laboratory Methods and Diagnosis*, Vols. I–III, Henry Kimpton, London.
GRELOU, G. (1958), Sahara, *Sci. et vie*, June 1958, Paris, pp. 18–49.
GRIFFITH, J. Q. and FARRIS, E. J. (1942), *The Rat in Laboratory Investigation*, J. B. Lippincott Co., London.
GULICK, A. (1937), *Amer. J. Physiol.*, 119: 322.
GUNN, D. L. (1942), Body temperature of poikilotherms, *Biol. Rev.*, 17: 293–314.
HALDANE, J. S. and GRAHAM, J. I. (1935), *Methods of Air Analysis*, Charles Griffin and Company, Ltd., London.
HARDY, J. D. and DuBOIS, E. F. (1938), *J. Nutr.*, 15: 461.
HART, J. S. (1956), Seasonal changes in insulation of the fur, *Canad. J. Zool.*, 34: 53–57.
HART, J. S. (1957), Climatic and temperature-induced changes in the energetics of homotherms, *Rev. canad. Biol.*, 16, No. 2: 133–174.
HART, J. S. and HEROUX, O. A. (1953), A comparison of some seasonal and temperature-induced changes in *Peromyscus*: cold resistance, metabolism and pelage insulation, *Canad. J. Zool.*, 31: 528.
HASSANEIN BEY, A. M. (1925), *The Lost Oases*, The Century Co., New York.
HATT, R. T. (1932), The vertebral columns of richochetal rodents, *Bull. Amer. Mus. nat. Hist.*, 63 (6): 599–738.
HAWK, P. B., OSER, B. L. and SUMMERSON, W. E. (1947), *Practical Physiological Chemistry*, J. and A. Churchill, Ltd., London.
HEDIGER, H. (1953), *Les animaux sauvages en captivité*, Payot, Paris.
HEIM DE BALSAC, H. (1936), Biogéographie des mammifères et des oiseaux de l'Afrique du Nord, *Bull. biol.*, Suppl. XXI, p. 447.
HEMINGWAY, A. (1945), Physiological effects of heat and cold, *Annu. Rev. Physiol.*, 7: 163–180.
HERRINGTON, L. P. (1940), The heat regulation of small laboratory animals at various environmental temperatures, *Amer. J. Physiol.*, 129: 123–139.
HILLPACH, W. (1944), *Géopsyché*, Payot, Paris, p. 347.
HIMWICH, H. E. (1945), Energy metabolism, *Annu. Rev. Physiol.*, 7: 181–200.
HORST, K., MENDEL, L. B. and BENEDICT, F. G. (1934), The effects of some external factors upon the metabolism of the rat, *J. Nutr.*, 7: 277–301.
HUNTINGTON, E. (1945), *Mainsprings of Civilization*, John Wiley and Sons, Inc., New York, p. 660.
INNES, W. (1932), Nos mammifères rongeurs, *Bull. Inst. Egypt.*, 14: 1–61.
IRVING, L. (1934), On the ability of warm-blooded animals to survive breathing, *Sci. Mon., N.Y.*, p. 422.
JACQUOT, R. and MAYER, A. (1926), *Ann. Physiol. Physicochim. biol.*, 11: 153.
JARVIS, C. S. (1936), *Three Deserts*, John Murray, Ltd., London, p. 290.
JARVIS, C. S. (1943), *Heresies and Humours*, Country Life Ltd., London, p. 176.
JOLEAUD, L. (1938), Paléogéographie du Sahara: histoire de la formation d'un désert, in *La vie dans la région désertique nord-tropicale de l'ancien monde*, Paul Lechevalier, Paris, pp. 21–47 (Société de Biogéographie VI).
KACHKAROV, D. N. and KOROVINE, K. P. (1942), *La vie dans les déserts*, Payot, Paris, p. 360.

BIBLIOGRAPHY

KANITZ, O. (1925), Body temperatures of animals, *Tabul. biol.*, Hague, 2: 9–14.
KAYSER, Ch. (1930), L'émission d'eau et le rapport $H_2O:O_2$ chez quelques espèces homéothermes et en cours de croissance, *Ann. Physiol. Physicochim. biol.*, 6: 721–744.
KAYSER, Ch. (1937), Variations du quotient respiratoire en fonction de la température du milieu chez le rat, le pigeon et le cobaye, *C.R. Soc. Biol.*, Paris, 126: 1219–1222.
KAYSER, Ch. (1940), Echanges respiratoires des hibernants, *Ann. Physiol. Physicochim. biol.*, 15 (5): 1087–1218 and 16 (1): 3–68.
KAYSER, Ch. (1954), Thermorégulation et métabolisme de l'eau, *Arch. Sci. physiol.*, 8: 245–264.
KAYSER, Ch. and DONTCHEFF, L. (1941), *Bull. Soc. Chim. biol.*, Paris, 23: 1229–1246.
KENDEIGH, S. C. (1934), The role of environment in the life of birds, *Ecol. Monogr.*, 4: 229.
KENDEIGH, S. C. (1945), Body temperature of small mammals, *J. Mammal.*, 26: 86–87.
KILLIAN, C. and FEHER, D. (1938), Le rôle et l'importance de l'exploration microbiologique des sols Sahariens, in AUFRÈRE, L. *et al.*, *La vie dans la région désertique nord-tropicale de l'ancien monde*, Paul Lechevalier, Paris, p. 105 (Société de Biogéographie VI).
KLEIBER, M. (1947), Body size and metabolic rate, *Physiol. Rev.*, 27: 511–541.
KUNO, Y. (1934), *The Physiology of Human Perspiration*, J. and A. Churchill, Ltd., London.
KUNO, Y. (1956), *Human Perspiration*, Charles C. Thomas, Springfield, Ill., p. 416.
LADELL, W. S. S. (1953), The physiology of life and work in high ambient temperatures, *Proc. Int. Symp. Desert Res.*, Jerusalem, p. 187–204.
LADELL, W. S. S., WATERLOW, J. C. and HUDSON, M. F. (1944), *Lancet*, ii: 491.
LAURENS, H. (1928), *Physiol. Rev.*, 8: 1.
LE BRETON, E. (1926), *Ann. Physiol. Physicochim. biol.*, 2: 606.
LE BRETON, E. and SCHAEFFER, G. (1923), *Recherches de physiologie générale sur la détermination de la masse protoplasmique active, et le rapport nucléoplasmatique*, Masson et Cie, Paris.
LE DANOIS, E. (1950), *Le rythme des climats dans l'histoire de la Terre de l'humanité*, Payot, Paris, p. 204.
LEE, D. H. K. (1940), *A basis for the study of man's reaction to tropical climates*, University of Queensland, Brisbane, Australia.
LEFÈVRE, J. (1911), *Chaleur animale et bioénergétique*, Masson et Cie, Paris, p. 1107.
LHOTE, H. (1937), *Le Sahara—désert mystérieux*, Editions Bourrelier, Paris, p. 127.
MACPHERSON, R. K. (1960), *Physiological responses to hot environments*, Medical Research Council Special Report Series No. 298, Her Majesty's Stationery Office, p. 323.
MANCHLE, W. and HATCH, T. F. (1947), *Physiol. Rev.*, 27: 200.
MARKHAM, S. F. (1944), *Climate and Energy of Nations*, Oxford University Press, New York.
MASON, E. D. and BENEDICT, F. G. (1934), *Amer. J. Physiol.*, 108: 377.
MASON, M. H. (1936), *The Paradise of Fools*, Hodder and Stoughton, London.
MAYER, A. and NICHITA, G. (1929), *Ann. Physiol. Physicochim. biol.*, 5: 609 and 774–841.
MEINERTZHAGEN, R. (1930), *Nicoll's Birds of Egypt*, Hugh Rees Ltd., London, Vols. I and II, pp. 348 and 700.
Meteorological Report for the years 1945–1957, Ministry of War and Marine, Egypt, Minerbo Press, Cairo, 1950.
MIGAHID, A. M. and ABDEL RAHMAN, A. A. (1953), Studies in the water economy of Egyptian desert plants, *Bull. Inst. Désert Egypt.*, 3, No. 1: 5–92.
MILLER, A. A. (1953), *Climatology*, Methuen and Co., London, p. 318.
MILLS, C. A. (1942), *Climate Makes the Man*, Harper and Brothers, New York and London, p. 320.
MINETT, F. C. and SEN, S. (1945), Rectal temperature of certain animals at rest, *Indian J. vet. Sci.*, 15, Part I: 63–78.

MITWALLY, M. (1952), History of the relations between the Egyptian oases, the Libyan Desert and the Nile Valley, *Bull. Inst. Désert Egypt.*, 2, No. 1: 115–131.

MITWALLY, M. (1953), Physiographic features of the Libyan Desert, *Bull. Inst. Désert Egypt.*, 3, No. 1: 147–163.

MOHAMED AWAD (1960), Settlement of nomadic tribal groups in the middle East, *Bull. Fac. Arts, Univ. Alexandria*, 14: 1–38.

MONOD, T. (1953), Exposé liminaire pour la section biologique, *Proc. Int. Symp. Desert Res., Jerusalem*, pp. 43–88.

MONTAGNE, R. (1947), *La civilisation du désert*, Librairie Hachette, Paris, p. 267.

MONTASIR, A. H. (1942), Soil structure in relation to plants at Maryut, *Bull. Inst. Egypt.*, 25.

MONTASIR, A. H. and SHAFEY, M. (1951), Studies on the autecology of *Fagonia arabica*, *Bull. Inst. Fouad ler Désert*, 1, No. 1: 55–75.

MORRISON, P. R. (1948), Oxygen consumption in several small wild mammals, *J. cell. comp. Physiol.*, 31, No. 1: 69–96.

MORRISON, P. R. (1948), Oxygen consumption in several mammals under basal conditions, *J. cell. comp. Physiol.*, 31, No. 3: 281–291.

MORRISON, P. R. and RYSER, F. A. (1952), Weight and body temperature in mammals, *Science*, 116, No. 3009: 231–232.

MUSIL, A. (1928), *The manners and customs of the Rwala bedouin*, American Geographical Society, New York.

NEWBURGH, L. H. (1949), *Physiology of Heat Regulation and the Science of Clothing*, W. B. Saunders Company, Philadelphia and London, p. 457.

NEWBURGH, L. H. and JOHNSTON, M. W. (1942), Evaporation in temperature regulation, *Physiol. Rev.*, 22: 1–16.

NIELSEN, M. (1938), *Skand. Arch. Physiol.*, 79: 193.

OGLE, C. and MILLS, C. A. (1933), Animal adaptation to environmental temperature conditions, *Amer. J. Physiol.*, 103: 606.

PAGÉ, E. (1957), Body composition and fat deposition in rats acclimated to cold, *Rev. canad. Biol.*, 16, No. 2: 269–278.

PETERS, J. P. (1950), Sodium, water and edema, *J. Mt. Sinai Hosp.*, 17: 159–175.

PIÉRY, M. et al. (1934), *Traité de Climatologie Biologique et Médicale*, Masson et Cie, Paris, Vols. I, II, III, pp. 904, 1800, 2664, resp.; see also GUIART, J., Zooclimatologie (climats et espèces animales) in Vol. I, p. 519.

POPOVA, T. V. (1946), Sur le mécanisme central de la thermorégulation physique, *Sechenov J. Physiol.*, 32, No. 5: 627–634.

POUQUET, J. (1951), *Les déserts*, Presses Universitaires de France, Paris, p. 124.

PRECHT, H., CHRISTOPHERSEN, J. and HENSEL, H. (1955), *Temperatur und Leben*, Springer-Verlag, Germany, p. 514.

PRENANT, M. (1934), Adaptation, écologie et biocœnotique. Actualités scientifiques et industrielles, in *Exposés de Biologie Ecologique*, Herman et Cie, Paris, p. 59.

PROSSER, C. L. et al. (1952), *Comparative Animal Physiology*, W. B. Saunders Company, Philadelphia and London, p. 888.

PROUTY, L. R. (1949), Heat loss and heat production of cats at different environmental temperatures, *Fed. Proc.*, 8: 128.

PRZIBRAM, H. (1923), *Temperatur und Temperatoren im Tierreiche*, Vienna.

RANDALL, W. C. (1943), *Amer. J. Physiol.*, 139: 56–63.

Report on the meteorological observations made at the Abbassia Observatory, Cairo, during the year 1900 together with the Alexandria mean values from the observations of the previous 10 years, The Survey Department, Public Works Ministry, National Printing Dept., Cairo, 1902.

ROBINSON, A. E. (1936), *The Camel in Antiquity*, McCorquodale and Co. (reprinted from *Sudan Notes*, 19, Part I).

BIBLIOGRAPHY

ROBINSON, S. (1949), Physiological adjustments to heat, in *Physiology of Heat Regulation*, edited by NEWBURGH, L. E., W. B. Saunders Company, Philadelphia and London, Chapter 5, pp. 193–231.

ROCHAIX, A. (1934), Adaptation aux climats, in *Traité de Climatologie Biologique et Médicale*, by M. PIÈRY, Masson et Cie., Paris, p. 1097.

ROCHE, J. (1926), *Arch. int. Physiol.*, 26: 5.

ROMER, A. S. (1954), *Man and the Vertebrates*, Pelican Books (A 303), Vol. I, p. 198, and Vol. II, p. 437.

RUBNER, M. (1902), *Die Gesetze des Energieverbrauchs bei der Ernährung*, F. Dieticke, Leipzig and Vienna.

RUSSELL, E. S. (1949), *Le comportement des animaux*, Payot, Paris, p. 232.

SANDERSON, I. T., *Living Mammals of the World*, Hanover House, Garden City, New York, p. 146.

SCHAEFFER, G. and POLAK, A. (1938), *C. R. Soc. Biol.*, séance du II mars: 127.

SCHMIDT-NIELSEN, B. (1949), Reduced evaporative water loss from the lungs of certain desert mammals, *Fed. Proc.*, 8: 139–140.

SCHMIDT-NIELSEN, B. (1954), Water conservation in small desert rodents, in *Biology of Deserts*, Institute of Biology, London, pp. 173–181.

SCHMIDT-NIELSEN, B. (1958), The resourcefulness of Nature in physiological adaptation to the environment, *Physiologist*, 1, No. 2.

SCHMIDT-NIELSEN, B. and SCHMIDT-NIELSEN, K. (1949), The water economy of desert mammals, *Sci. Mon.*, 69: 180–185.

SCHMIDT-NIELSEN, B. and SCHMIDT-NIELSEN, K. (1950), Evaporative water loss in desert rodents in their natural habitat, *Ecology*, 31: 75–85.

SCHMIDT-NIELSEN, B. and SCHMIDT-NIELSEN, K. (1950), Pulmonary water loss in desert rodents, *Amer. J. Physiol.*, 162: 31–36.

SCHMIDT-NIELSEN, B. and SCHMIDT-NIELSEN, K. (1951), A complete account of water metabolism in kangaroo rats and an experimental verification, *J. cell. comp. Physiol.*, 38: 165–181.

SCHMIDT-NIELSEN, B., SCHMIDT-NIELSEN, K., BROKAW, A. and SCHNEIDERMAN, H. (1948), Water conservation in desert rodents, *J. cell. comp. Physiol.*, 32, No. 3: 331–380.

SCHMIDT-NIELSEN, K. (1954), Heat regulation in small and large desert mammals, *Biology of Deserts*, Institute of Biology, London, pp. 182–187.

SCHMIDT-NIELSEN, K. (1959), The physiology of the camel, *Sci. Amer.*, December 1959, p. 140.

SCHMIDT-NIELSEN, K. and SCHMIDT-NIELSEN, B. (1952), Water metabolism of desert mammals, *Physiol. Rev.*, 32, No. 2: 135.

SCHOLANDER, P. F., HOCK, R., WALTERS, V. and IRVING, L. (1950), Adaptation to cold in arctic and tropical mammals and birds in relation to body temperature, insulation and basal metabolic rate, *Biol. Bull.*, 99: 259.

SCHOLANDER, P. F., HOCK, R., WALTERS, V., JOHNSON, F. and IRVING, L. (1950), Heat regulation in some arctic and tropical mammals and birds, *Biol. Bull.*, 99: 237.

SCHOLANDER, P. F., WALTERS, V., HOCK, R. and IRVING, L. (1950), Body insulation of some arctic and tropical mammals and birds, *Biol. Bull.*, 99: 225.

SCOTT, J. C. and BAZETT, H. C. (1941), Temperature regulation, *Annu. Rev. Physiol.*, 3: 107.

SEALANDER, J. A. (1953), Body temperature of white-footed mice in relation to environmental temperature and heat and cold stress, *Biol. Bull.*, 104: 87.

SEARS, P. B. (1949), *Deserts on the March*, Routledge and Kegan, Ltd., London, p. 181.

BIBLIOGRAPHY

SELLERS, E. A. (1957), Adaptive and related phenomena in rats exposed to cold, *Rev. canad. Biol.*, 16, No. 2: 175–188.

SERGENT, E. (1954), Le peuplement humain du Sahara, in *Biology of Deserts*, Institute of Biology, London.

SETZER, H. W. (1958), The jerboas of Egypt, *J. Egypt. publ. Hlth. Ass.*, 33, No. 3: 87–94.

SHAFEI, A. (1952), Mareotis Lake, its past history and its future development, *Bull. Inst. Désert, Egypt.* 2, No. 1: 71–101.

SHEARD, C. (1950), Temperature of skin thermoregulation of the body, in *Medical Physics of Otto Glasser*, The Year Book Publishers, Inc., Chicago, Ill., pp. 1523–1555.

SHERWOOD, T. C. (1936), The relation of season, sex, and weight to the basal metabolism of the albino rat, *J. Nutr.*, 12: 223–236.

SHOTTON, F. W. (1954), The availability of underground water in hot deserts, in *Biology of Deserts*, Institute of Biology, London.

SHREVE, F. (1934), The problems of the desert, Desert Laboratory of the Carnegie Institution of Washington, *Sci. Mon.*, March 1934, p. 199.

SHREVE, F. (1936), The plant life of the Sonoran Desert, Desert Laboratory of the Carnegie Institution of Washington, *Sci. Mon.*, March 1936, p. 195.

SIPLE, P. A. (1949), Clothing and climate, in *Physiology of Heat Regulation and the Science of Clothing*, W. B. Saunders Company, London and Philadelphia, p. 389.

SMITH, H. W. (1956), *Principles of Renal Physiology*, Oxford University Press, N.Y.

SORRE, M. (1934), Les climats de la terre moins l'Europe, in *Traité de la Climatologie Biologique et Médicale*, Masson et Cie, Paris, pp. 457, 458.

SORRE, M. (1934), Le climat et les sociétés humaines, in *Traité de Climatologie Biologique et Médicale*, Masson et Cie, Paris, p. 565.

SORRE, M. (1951), Les fondements de la géographie humaine, in *Les Fondements Biologiques*, Vol. 1, Librairie Armand Colin, Paris, p. 448.

SPECTOR, W. S. (1956), *Handbook of Biological Data*, W. B. Saunders Co., Philadelphia and London.

STEEL, J. H. (1890), *A Manual of the Diseases of the Camel*, Asylum Press, Madras, p. 206.

STRAUSS, M. B. (1957), *Body Water in Man*, J. and A. Churchill Ltd., London.

SUNDSTROEM, E. S. (1927), *Physiol. Rev.*, 7: 320.

SUTTON, L. J. (1946), *The Climate of Egypt*, R. Schindler, Cairo, p. 50.

TADROS, T. M. (1953), A phytosociological study of halophilous communities from Mareotis (Egypt), *Vegetatio*, Haag, 4, 15 III, Fasc. 2.

TANON, L. and NEVEU, R. (1934), Climatopathologie désertique, in *Traité de Climatologie Biologique et Médicale*, Masson et Cie, Paris, p. 1140.

TAYLOR, G. (1927), *Environment and Race*, Milford, London.

TERMIER, H. and TERMIER, G. (1952), *Initiation à la Paléontologie*, Collection Armand Colin, Paris, No. 273–274, pp. 199, 211.

TERROINE, N. and ROCHE, J. (1925), *Arch. int. Physiol.*, 24: 356.

THIBAULT, O. (1949), Les facteurs hormonaux de la régulation chimique de la température des homéothermes, *Rev. canad. Biol.*, 8, No. 1: 136.

THOMPSON, D'ARCY W. (1942), *On Growth and Form*, Cambridge University Press, Cambridge.

UNSTEAD, J. F. (1948), *A Systematic Regional Geography. A World Survey from the Human Aspect*, Vol. III, University of London Press, Ltd., London, p. 451.

VIAUD, G. (1955), *Arch. Sci. physiol.*, 9 (4): 35–47.

VORHIES, C. T. (1945), *Water requirements of desert animals in the southwest*, University of Arizona Agricultural Experimental Station, Technical Bulletin No. 107, pp. 487–525.

WASSIF, K. (1953), On a collection of mammals from Northern Sinai, *Bull. Inst. Désert Egypt.*, 3, No. 1: 107–118.

WEINER, J. S. (1954), Human adaptability to hot conditions of deserts, in *Biology of Deserts*, Institute of Biology, London.

BIBLIOGRAPHY

WELSH, J. H. (1938), *Quart. Rev. Biol.*, 13: 123.

WILLIAMS, C. B. (1923), *A short bioclimatic study in the Egyptian desert*, Ministry of Agriculture, Egypt, Technical and Scientific Service, Bulletin No. 29, Government Press, Cairo, p. 30.

WILLIAMS, C. B. (1924), *Bioclimatic observations in the Egyptian desert in March, 1923*, Ministry of Agriculture, Egypt, Technical and Scientific Service, Bulletin No. 37, Government Press, Cairo, p. 18.

WILLIAMS, C. B. (1924), *A third bioclimatic study in the Egyptian desert*, Ministry of Agriculture, Egypt, Technical and Scientific Service, Bulletin No. 50, Government Press, Cairo, p. 32.

WILLIAMS, C. B. (1954), Some bioclimatic observations in the Egyptian desert, *Biology of Deserts*, Institute of Biology, London, pp. 18–27.

WILSON, E. B. (1952), *An Introduction to Scientific Research*, McGraw-Hill Book Co., New York.

WINSLOW, C.-E. A. and HERRINGTON, L. P. (1949), *Temperature and Human Life*, Princeton University Press, Princeton, New Jersey, p. 272.

WORDEN, A. N. (1947), *The UFAW Handbook on the Care and Management of Laboratory Animals*, Baillière Tindall and Cox, London, p. 368.

World Almanac and Book of Facts—1956, World-Telegram and The Sun publication, New York.

WULSIN, F. R. (1949), Adaptations to climate among non-European peoples, in *Physiology of Heat Regulation and the Science of Clothing*, W. B. Saunders Company, Philadelphia, p. 3.

YEATES, N. T. M. (1955), Photoperiodicity in cattle. Seasonal changes in coat, character and their importance in heat regulation, *Aust. J. agric. Res.*, 6: 891.

INDEX

Acclimatization
  adrenal cortex, and, 88
  body temperature, and, 55
  desert heat, to, 87–88
  physiological phenomenon, as a, 76
Acetylcholine, 85
Activity cycle
  animal and body temperature, 50
  effect of nutrition, 47–49
  24-hour spontaneous, 47–49
Actograph, technique, 111, 112
Adaptations
  biota to desert, 1–12
  desert fauna, 8, 9
  desert flora, 9–12
  ecological, 8, 13
  ethological (behavioural), 8, 13
  genetic, 8, 13
  morphological (structural), 10, 13
  physiological, 10, 13
  thermoregulatory, 56
Adda gazelles, 11
Addax, 10, 11
Adrar, 6
Adrenal cortex, 88
Adrenalin, 59
Adrenalinemia, 58
Aestivation, *see* Estivation
Africa, 1, 7, 9, 14, 77
Air
  dryness of, 2
  hot currents, 82
  movement rate, 23, 27
Alcohol, 85
Aldosterone, 85
Alexandria, 24
Algeria, 4, 15
Allactaga tetradactyla, 15, 16
Almaza, 22, 23
America, 1, 7
Animals
  hoofed, 7
  species, 50
Ant-bears, 11
Antelopes, 9, 11
Antidiuretic hormone (ADH), 84, 85
  dehydration, and, 86
Anuria, 89
Appetite, 89
Arabia, 1, 15, 77

Arabs, desert, 77
Argentina, 1, 77
Aridity, 2, 3, 23
Ariel gazelles, 11
Arizona, 28
Ascopharynx, 11
Asia, 1, 7, 14, 15
Aswan, 15
Atropine, 85
Australia, 1, 7, 77

Basal Metabolism, 57, 58
  jerboa's, 12, 60–62
  percentage of thermolysis, 65
  rat's, 12, 60–62
Bedouin, 21, 74, 77
  Rwala, 78
Berbers, 77
Betpak-Dala, 4
Bilma, 6
Biocenosis, desert, 7
Biota, desert, 7–9
Birds, 7, 10
Biskra (Algeria), 5
Blood
  circulation, 82
  plasma, 83
  volume, 82
  see Plasma
Body
  active surface, 64
  attitude, 59, 64
  composition of electrolytes, 83
  concentration of electrolytes, 83
  fat content, 45–47, 110–111
  feathers, 64
  fluid movement under thermal stress, 84
  fluids, 83
  fur, 64
  osmotic pressure, 83
  'physiological service', 57
  reactions to ambient heat, 82
  size and evaporative water loss, 82
  sodium chloride content, 84
  temperature
    cycle of activity, 50
    daily cycle in rats, 50
    dehydration, 89
    heat balance, 80–81

INDEX

Body—*continued*
   heat stroke, 89
   jerboas and white rats, 12, 50–56, 132–133
   method of taking, 104–105
   regulation, 50–56
   sweating, 86
   thermal neutrality, 55
   zone of regulation, 75, 135–136
   zone of regulation in jerboas on wet and dry diets, 134–135
  waste products, 84
  water content, 45–47
   effect of diet on, 42, 45–47
   method of measuring, 110–111
  weight
   dehydration, 89
   effect of alternating wet and dry diets, 40–41, 121
   effect of dry diet, 40, 119
   effect of wet and dry diets, 42–44, 121–131
   jerboas, 19, 113
   variations during growth and after maturity, 37–39, 116–118
   variations on dry diet, 40–41, 120
Borg El Arab, 21
Breathing rate and dehydration, 89
Bullae tympani, inflated, 10
Burrows, 9, 12
  jerboa's, 24–25
   burrow-digging, 24
   temperature and humidity, 28–29
Bustards, 11

Cairo, 23
California, Death Valley, 4
Calorimetry, indirect, 105–107
Camels
  heat tolerance and water economy, 11
  nomadic people, and, 78
  publications on, 11
Carnivora, 7, 9, 10
Cats, 64
Cenozoic era, 14
Central Asia, 15
Chile, 1, 77
China, 77
Climate
  definition, 3
  elements and factors, 3
  hot desert
   above ground, 26
   beneficial effects, 76
   characteristics, 2–6
   coastal zone west of Alexandria, 21–23
   human race, and, 76–77
   stresses on man's body, 80
  see Aridity, Macroclimate
Clothing
  desert, in, 78, 82, 89
  sweating rates, 87
Cold
  exposure to, 84
  zone of, and thermogenesis, 59
Colorado, 1, 77
Coloration, desert animals, 9–10
Conduction, 59, 81, 82
Convection, 59, 81, 82
Cutaneous circulation, 82

Dehydration
  antidiuretic hormone, 86
  anuria, 89
  body temperature, 88–90
  body weight, 89
  breathing rate, 89
  deafness, delirium, and dimness of vision, 89
  endurance to, 88
  evaporative water loss, 82
  individual differences, 88
  lack of appetite, 89
  pulse rate, 89
  swollen tongue, 89
  thirst, 89
  working efficiency, 86
'Dehydration exhaustion', 89
'Dehydration reaction', 88
'Dehydration, voluntary', 80
Desert
  areas, 2
  biology, 90
  biota, 7–8
  civilization, 77
  climate
   acclimatization, 86
   characteristics, 2
   definition, 3
   effects on man, 90
   man's salt needs, 90
   salinity of water, 86
  dwellers, 77
  fauna, 9
  fertilization, 90
  flora, 8–9
  hazards, 88–90

INDEX

Desert—*continued*
　macroclimate, 2–6
　mammals, 65
　resources, 90
　rodents, 65
　way of life, 78
Desoxycorticosterone acetate (DCA), 88
Dew, 2, 6
Diet, wet and dry, effect on
　body temperature, 51–52, 132–133
　body weight, 40–41, 119–121
　energy metabolism, 60–63, 137–142
　excretion, 42–45, 121–131
　heat balance, 72–75
　insensible perspiration and evaporative water loss, 64–71, 143–147
　zone of body temperature regulation, 52–55, 134–136
　24-hour activity cycle, 47–48
　see Nutrition
Dipodidae, 14, 15
Dipodillus, 11
Dipodomys (kangaroo rat), mechanisms of water conservation, 11–12
　see Kangaroo rat
Dipus, 11, 15
Dipus aegyptius, *see* Jerboa
Diuresis, 85
Dogs, 64
Dorcas gazelles, 11
Dormancy, 10
Drinking in desert, 89

Earth Areas, desert, fertile, steppes, 1
Ecoclimate, jerboa's, 24–29
Ecology, 7
Egypt, 15, 24
　advent of desert conditions, 15
　fauna and flora of oases, 15
Egyptian Desert (near Cairo), 5, 24
Egyptian Western Desert (littoral zone), 21–24, 30
El Agheila, 16
El Alamein, 21
El Gharbanyat, 21
Energy metabolism, 57–63
　jerboa's and rat's, 59–63, 137–142
Eocene epoch, 14
Equator, 1, 7
Equatorial forest, 7
Estivation, 25, 63
Europe, 1, 9, 14, 15
Evaporation
　Libyan desert, in, 5
　rate, 2, 5, 23, 27
　ratio to rainfall, 5
Evaporative water loss (cutaneous and respiratory)
　Dipodomys, 12
　jerboas and rats, 64–71, 75, 143–147
　latent heat of vaporization, 59, 63, 64–71, 143–147
　man, 81, 82
　thermolysis, 58–59, 64–71
　see Insensible perspiration
Evolution, 74, 94
　jerboa's thermoregulation, 56
　thermoregulatory mechanisms, 50
Excretion, effect of diet on, 42–45, 48–49, 122–131
　see Feces, Urine
Extracellular fluid, 83, 86, 88

Fat, content of animal body, 45–47, 110–111
　insulating power, 11
Fatty acids, oxidation of, 59
Fauna, 14
　desert, 9, 62
　Egypt's oases, 15
Feces
　camel's, 11
　jerboa's and rat's, 13, 42–45, 48–49, 62, 122–132
　man's, 81
　see Excretion, 81
Filipinos, 88
Flora
　desert, 8–9, 24, 62
　Egypt's oases, of, 15
Food
　heat production, 82
　scarcity, 31, 42, 74
Foxes, 9, 11
Fur, 10, 20

Gastrointestinal upsets and dehydration, 89
Gazelles (adda and ariel), 9, 11
Gerbillus, 11
Gerbils, 11
Ghardaia, Algeria, 4, 5
Glomerular filtration, 84
Glucosides, 62
Glycogenolysis, 59
Goats, 78
Gobi desert, 1, 11, 77

INDEX

Growth, comparative, of jerboas and rats, 37–41, 116–117

HAIR, insulating power, 11
Hamsters, 65
Heart, rate, 82
  systolic volume, 82
Heat
  balance, jerboa's and rat's, 12, 62, 72–75, 108–109
    man's, 80–82
  conservation, jerboa's, 34
  cramp, 90
  environmental, 1, 31
  exhaustion, 80, 90
  exposure to, man's, 84
  gain, man's, 80, 82–83
  load, in the desert, man's, 82–83
  loss, jerboa's and rat's, 64–71, 143–147
    man's, 80, 81–83
  of evaporation, latent, 65
  production, man's, 82–83
  protection against, man's, 82–83
  routes of elimination, man's, 81
  stroke, 80, 89
  see Temperature
Hemoconcentration, 84, 89
Hemodilution, 84
Herbivora, 9–10
Heteromyidae, 42, 65
Heterothermic animals, 50
Hibernation, 10, 25, 62
Homotherms, 50
  hypothermia, 55
  thermoregulation, 55, 58
Hormones, 59
Human race
  adaptation to desert climate, 76–90
  coloured (black), 76–77
  white, infant mortality, 77
  women, 77
Humidity, relative, 2, 4, 5, 22–23, 27, 28, 59, 82
Hyenas, 11
Hyperthermia, 41
Hypothalamico-hypophyseal system, 85
Hypothalamus, 82, 85
Hystricidae, 15

INANITION, 45
Insects, 9, 10
Insensible perspiration and evaporative water loss

jerboas and rats, 64–71, 75, 143–145
  man, 81
  method of studying, 107–108
Insolation, 1, 3–4, 22
Intermedine, 59
Interstitial fluid, 83
Intestinal tract, 45
Intestines, ballooned, 41
Intracellular fluid, 83
  shift of water, 86
Iraq, 15

JACKALS, 9
Jaculidae, 14, 15
Jaculus, 11, 14
Jaculus jaculus, 15–17, 30, 94
Jaculus orientalis, see Jerboa
Japanese, 88
Jerboa (*Dipus aegyptius* or *Jaculus orientalis*)
  activity cycle, 24-hour, 47–48
  anatomical features, 16–18
  basal metabolism, 12, 60–62
  body development, 38
  body parts and dimensions, 16, 19, 114
  body temperature, 12, 50–52, 132–133
    thermal neutrality, 53, 55
    zone of regulation, 52–56, 134–135
  body weight
    during growth and after maturity, 118
    effect of diet on, 39–40, 119–121
    sex, 39, 113
  breeding season (reproduction), 30
  burrows, 17, 24–25
    evaporative water loss, 29
    heat regulation, 29
    microclimate, 27–29
    temperature and humidity, 115
  desert species, as a, 14–20, 30
  distribution, 15–16
  ecoclimate, 24–29
  enemies, 30
  energy metabolism, 57–63, 137–139
  excretion (*see below*, nutrition and excretion)
  feces excretion, 42–45, 48–49, 122–131
  food, 29
    consumption, 44
  fur, 20
  growth, 37–41, 116–117
  habitat, 21–29
    climate, 21
    desert environment, 21–24

INDEX

Jerboa—*continued*
  physiography, 21
  rainfall, 22
  temperature and humidity, 22
  vegetation, 24
  heat balance, 12, 72–75
  heat production, 61–63
  in the laboratory, 32–37
    behaviour, 35
    crouching position, 34
    ecoclimate, preferred, 33
    cage, 33
    pseudo-burrow, 32–33
    identification, 35–36
    nutrition, 32
    reproduction, 36
    rest and sleep, 34
  insensible perspiration and evaporative water loss, 12, 53, 64–71, 143–147
  kidneys, purging, 41
  locomotion, 17, 29
  longevity, 38
  morphological observations, 19–20
    mammary glands, 20
  muscular relaxation, 51
  nocturnal life, 12, 26–27
  nutrition and excretion, 42–45, 48–49, 122–127, 130–131
  origin, 14–15
  physiological and genetic adaptations, 55
  renal elimination, 40
  salivation, 55
  sex identification, 36
  sexual maturity, 38
  skin area, 19
  sleep, deep, 51
    'estivation reflexes', 63
    heat stress, 55–56
    state of lethargy, 62–63
  social habits, 30
  thermal neutrality, 60–61
  thermoregulation, 72–75
  urea elimination, 41
    concentration, 44
  urinary system, 41
  urine concentration, 44
  urine excretion, 42–45, 48–49, 122–131
  urine reaction, pH, 44, 122–129
  water balance, 12, 72–75
  zone of physical regulation (thermolysis), 61–63
Jerboas, 9, 11, 14

KALAHARI, desert, 1, 77
  bushmen's fat deposit, 76
Kangaroo rat, 11–12, 28, 42, 64–65
Kara Kum, 4
Kasalinsk, Turkestan, 4
Khamsins, 22
Kidneys
  excretion of salt and water, 62, 84
    urine, 86
    waste products, 41, 84
  hormonal and neurogenic regulation, 84
  maximum concentration of urea, 41
  regulation of body fluids and blood, 84

LEPORIDAE, 15
Lethargy, state of, 63
Libya, 10, 15
Libyan desert, 4, 5, 10, 15, 77
Light, 2
Lions, 11
Liver, anemic and granulated, 41
Lizards, 7, 10
Lymph channels, 86

MACROCLIMATE, desert, 2–6
  see Climate *and* Desert
Mammals, 10, 50
Man
  acclimatization to desert, 87–88
    adrenal cortex, 88
    improved heat regulation, 87
    individual differences in blood circulation, sweat formation, and salt conservation, 87–88
  acetylcholine, 85
  adaptation to desert climate, 76–90
  adrenal cortex, 88
  adrenocortical hormones, 88
  alcohol, 85
  aldosterone, salt-retaining factor, 85
  and the deserts, 77–79
    characteristics of desert dwellers, 77
    clothing in the desert, 78, 82
    'desert civilization', 77
    desert climatic elements, 80
    hot desert climate stresses, 80
    nomadic life, 78–79
      eating and drinking habits, 79
      tribal organization, 79
    shelter in the desert, 78
    way of life, 78
  antidiuretic hormone (ADH), 84–86
    inhibited release, 85

INDEX

Man–*continued*
  secreted by neurohypophyseal system, 84
    stimulated release, 85
  atropine, 85
  blood plasma, 83
    volume changes, 84
  body fluids, 83–85
    extracellular fluid, 83
    fluid movements, 84
    interstitial fluid, 83
      volume changes, 84
    intracellular fluid, 83
      volume changes, 84
    see below, Kidneys, Sodium chloride, Sweat, Water
  body heat and water exchanges, 81
  body heat balance, 80–83
  body heat elimination routes, 81
  body physiological adjustments, to environmental heat, 82
    to hot desert, 80
  body protection against environmental heat, 82
  body reactions to
    ambient heat and heat load, 82
    excess heat or cold, 84
    thermal stress and heat tolerance limits, 81, 87
  body temperature, 80–81, 86, 87
  body thermoregulation, 76
  climate
    and human race, 76–77
    beneficial effects, 76–77
    chronic effects, 90
    influence, 76–77
  dehydration, 83, 88–90
  dehydration exhaustion, 89
  'dehydration reaction', 88
  desert hazards, 88–90
    dehydration, 88–89
    dehydration exhaustion, 89
    heat cramp, 90
    heat exhaustion, 90
    heat stroke, 89
    'voluntary dehydration', 80, 89
  desoxycorticosterone acetate (DCA) and lowering of salt concentration, 88
  evaporation rate, 87
  evaporative water loss, 82
  feces, 81
  heat cramp, 90
  heat exhaustion, 90
  heat stroke, 89
  hemoconcentration, 84
  hemodilution, 84
  hypothalamico-hypophyseal system and release of ADH, 85
  hypothalamus and 'salt-excreting' hormone, 85
  kidneys, 83–86
    renal control of salt and water excretion, 84
    urinary response, 84
    see below, Sodium chloride, Sweat, Water
  lungs, 81
  nicotine and release of ADH, 85
  osmotic equilibrium, 84, 86
  osmotic pressure, 83
  osmotic regulation, 83–84
  osmotic work of sweat glands, 86
    see Body fluids, Sodium chloride, Sweat, Water
  pilocarpine and sweat stimulation, 85
  pituitary body, 86
  pulse, pressure, 87
    rate, 81, 87
  skin, 76, 81, 83, 86
    temperature, 82, 87
      critical to initiate sweating, 86
  sodium chloride (salt), 85–86
    appetite for, 84
    body content, 84
    concentration in sweat, 85–86
    depletion, 83
    excretion, 84
    mechanism of concentration in serum and extracellular fluid, 86
    mechanism of conservation, 84–85
  supraoptico-hypophyseal system and renal excretion of water, 85
  survival limits in desert, 88
  sweat
    centre, mental and thermal, 85
    glands, 85
      'fatigue', 87
      Filipinos, 88
      Japanese, 88
      osmotic work of hypotonic sweat secretion, 86
      Russians, 88
      tropics, natives of, 88
    rate in the desert, 86–87
      capacity for muscular effort, 86–87
      military activity, 87
      night work, 87

INDEX

Man—*continued*
  secretion
    at rest, 81
    inhibited by atropine, 85
    stimulated by acetylcholine and pilocarpine, 85
    with exercise, 81
  sodium chloride concentration, 85–86
    lowering by adrenocortical hormones, 88
  technology, 76
    air-conditioning, 76
    clothing, 76
    shelter, 76
  thirst mechanism, 84, 89
  urine, 81, 84, 86, 88
  'voluntary dehydration', 80, 89
  water and salt, balance, 83–85
    conservation mechanisms, 84–85
    excretion, 83–84
  water elimination routes, 81
  see Human race, Bedouin, Desert
Mareotis (west of Alexandria), 24
Marmots, 14
Maryut, 16, 21, 22, 28
Membrane permeability to solutes, 83
Meriones, 11
Meriones shawi, 35
Mersa Matruh, 16, 22–23
Metabolism
  activity, 73–74
  basal, 73–74
  energy, 57–63, 137–142
  water, 42, 65
Microclimate, 27–29, 34
  and desert life, 6–9
  see Climate *and* Desert
Microorganisms adapted to dryness, 6
Monkeys, 11
Morocco, 15
Morphological adaptations
  character, 10
  coloration, 10
Muridae, 15
Muscle spasm and heat cramp, 90
Muscular activity, 82
Myoxidae, 15

NAUSEA and dehydration, 89
Neurohypophyseal secretion of ADH, 84–85
Nicotine, 85
Night work in desert, 89
Nile valley, 15

Nocturnal, animals, 9
  jerboa, 12, 26–27
  life, 27
Nomads, 90
  behavioural adjustments, 78–79
  diet, 78–79
  eating and drinking habits, 79
  life, 78–79
  populations, 77
  see Bedouin, Man and the deserts
Noradrenalin, 59
North Africa, 15, 30
  littoral, 4
  neolithic epoch, 9
  winds, 4
Notomys, 11
Nubian desert, 88
Nutrition, effect of, on
  body weight, 39–41, 119–121
  energy metabolism, 60–63, 137–139
  excretion, 42–45, 122–131
  heat and water balance, 72–75
  insensible perspiration and evaporative water loss, 66–71, 143–144
  spontaneous activity, 47–48
  thermoregulation, 72–75
  water content of the body, 45–47
  see Diet
Nutrition and excretion, 42–45, 122–131
  method of studying, 109–110

OASES, 77
Oligocene grottoes, 14
Oran, Algeria, 5
Oryx, 11
Osmotic equilibrium of body fluids, 84, 86
Osmotic regulation, 83–84
Ostriches, 11
Oxygen consumption
  evaporative water loss, 69
  jerboas and white rats, 137–147
  method of studying, 105–109

PACHYUROMYS DUPRASI, 35
Palearctic desert, 17
  great, 11, 77
Paleocene, 14
Palestine, 4, 15
Paludism, 77
Panting, 85, *see* Polypnea
Paramys, 14
Pedetes, 11

165

Permian time, 7
Pigeon, 50
Pilocarpine, 85
Pilomotor reflexes, 59
Plants, desert, 6, 7–9
Plasma, regulation of osmotic pressure and volume, 84
Pleistocene, 15
Podoces panderi, 7
Poikilotherms, 50
Polypnea, thermal, 59, 63, 64
   see Panting
Population, world, 1
Porcupines, 11
Proteins, 62
Pulmonary ventilation, 59
Pulse rate and dehydration, 89
   and heat exhaustion, 90

QUATERNARY PERIOD, 14
   glacial period, 15

RABBITS, 64
Radiation, sky and ground, 2, 82
   solar, 3–4, 27, 31, 82
Radiation from body surface, 59, 81, 82
Rain, 1
Rainfall in hot deserts, 2, 6, 22–23
Rat, white (Wistar)
   activity cycle, 24-hour, 42, 47–48
   basal metabolism, 12, 60–62
   body development, 38
   body temperature, 50–52, 132
      thermal neutrality, 55
      zone of regulation, 52–56, 136
   energy metabolism, 57–63, 140–142
   excretion
      see below, nutrition and excretion
   feces, 44–45, 48–49, 130–131
   growth, 37–41, 116–117
   heat balance, 72–75
   heat production, 61–63
   in jerboa's company (cage), 35
   insensible perspiration and evaporative water loss, 64–71, 144–147
   longevity, 38
   nutrition and excretion, 41, 44–45, 48–49, 130–131
   sexual maturity, 38
   thermal neutrality, 60–61
   thermoregulation, 72–75
   urine, 44–45, 48–49, 130–131
   water balance, 72–75
   zone of physical regulation (thermolysis), 61–63

Rats, 14
Relaxation, 82
Renal elimination, 40
   see Kidneys
Reproduction in captivity, 36
Reptiles, 7, 10
Resistance
   climatic, 76
   heat, by jerboas, 63
   individual differences, 76
Respiration rate, 82
Respiratory exchanges, 60
   see Energy metabolism
Respiratory Quotient, 106, 137–145
Rhine Valley, 15
Rodentia, 14
Rodents, 7, 14, 64, 65
Russia, 15
Russians, 88
Rwala bedouin, 78
   see Bedouin *and* Nomads

SAHARA DESERT, 1, 4, 5, 6, 8, 77
Salivation, 55, 59, 61, 63–71, 85
Salt (sodium chloride)
   balance, man's, 80, 83–85
   conservation mechanisms, 84–85
   depletion, 80
   elimination in sweat, 80, 86
   excretion, 84–85
   physiological solution, 83
   'salt-excreting hormone', 85
   see Man, Sodium chloride, Kidneys, Sweat
Savannah, 7
Scirtomys tetradactylus, 15, 16
Sciuridae, 15
Sensitivity, individual differences, 76
   to climatic conditions, 76
Serpents, 7, 10
Serum, chloride concentration, 86, 88
   osmotically effective concentration, 83
Shade, 82
Sheep, 78
   mountain, 11
Shelter, 82
Shivering, 59
Simooms, 4
Simplicidentata, 14
Sinai, 15
Siroccos, 4
Skin
   area, jerboa's, 19
   capillaries, man's, 76

INDEX

Skin—*continued*
  chloride accumulation, man's, 86
  impervious (desert animals), 10
  sweat glands, man's, 76, 88
  temperature, man's, 82, 86, 87
    sweating, 86
  wetted area, 82
Sky, clouds, 2–3
Sleep, jerboa's
  deep (lethargic), 55, 63, 66, 71
  normal, 34
  see Estivation
Sodium chloride concentration, 85–86
  see Salt, Sweat, Urine
Soils, desert type, 2, 6
  microorganisms, 6
  temperatures, 2
Somaliland, 15
Sonoran desert, 8
South America, 14
Spalacidae, 15
Squirrels, arboreal, 14
Steatopygia, 76
Steppes, 1, 9, 17
Sudan, 15
Summary and conclusions, 91–95
Supraoptico-hypophyseal system, 85
Susliks, 10
Sweat, man's
  acclimatization, 87
  constituents, 85
  glands, man's
    active number, 88
    eccrine and apocrine, 85
    'fatigue', 87
    osmotic work, 86
    secretory nerves, 85
  heat loss through, 85
  hypotonic, 86
  sodium and chloride ions, 85
  vitamins in, 85
Sweating
  acetylcholine, and, 85
  air temperature, and, 86
  atropine, and, 85
  body temperature, and, 86
  camel's, 12
  daily loss of salt, and, 86
  desert climate, and, 86
  excessive, 86
  hypothalamus, and, 82
  man's, 85–87
  mechanisms of, 80
  pilocarpine, and, 85
  rate of, 86–87
  skin temperature, and, 86
  thermal and mental stimulation, 85
Syria, 4, 15

TAMANRASSET, 4
Technique, 99–112
  actograph, 111–112
  animals studied, 103–104
  body temperature, 104–105
  heat and water balance, 108–110
  indirect calorimetry, 105–107
  insensible perspiration and evaporative water loss, 107–108
  laboratory equipment, 99–103
  nutrition and excretion, 109–110
  water and fat content, 110–111
Tehennu, 77
Temperate forest, 7
Temperature, air, 4, 22–23, 27, 59, 82
  environmental, and
    body-temperature regulation, 52–56
    energy metabolism, 57–63
    insensible perspiration and evaporative water loss, 64–71
    physiological adjustments, 82
    skin temperature, 82
    sweating rate, 86
    thermal neutrality, 53, 55, 57–62, 64–70
    thermogenesis, 57, 59, 61
    thermolysis, 58–59, 64–71
    thermoregulation, 72–75
      chemical, 58–59, 61–63
      physical, 58–59, 61–63
  ground, 28
  soil surface, 4, 23
  sun, 28
  variations (diurnal, nocturnal, seasonal), 2
  zone of low (cold), 59, 62
  zone of high (hot), 62
  see Heat
Tents, 78
Tertiary period, 14
Thar desert, 1, 77
Thermal adaptation, 58
Thermal neutrality, 53, 55, 57–62, 64–70
  see Energy metabolism
Thermogenesis, 57, 59, 61
Thermolysis, 58–59, 64–71
  active surface, and, 64
  convection, conduction and evaporation, and, 64

Thermolysis–*continued*
  fur, feathers, and, 64
  polypnea, and, 64
  radiation, and, 64
  salivation, and, 55, 59, 61, 63–71, 85
  sweating, and, 64
  wetted skin area, and, 82
  wetting of body parts, and, 61, 66, 85
Thermoregulation, 41, 53, 55, 56, 64, 72–75
  body temperature, and, 75
  chemical regulation, 58–59, 61–63
  diet, and, 75
  energy metabolism, and, 75
  heat and water balance, 72–75
  insensible perspiration and evaporative water loss, and, 75
  physical regulation, 58–59, 61–63
Thirst mechanism and dehydration, 89
Thyroxine, 59
Thyroxinemia, 58
Tindouf, 6
Toes, fringed, 10
  reduced number, 10
Tongue, swollen, and dehydration, 89
Torpor, hibernal, 10
  see Hibernation
Tripoli, 15
Tropical forest, 7
Tropics
  chlorine concentration in sweat, 88
  sweat glands, 88
  sweating in, 87
Tropics of Cancer and Capricorn, 1
Tuaregs, 77
Tundra, 7
Tunis, 15
Turkestan desert, 1, 77

Unesco, 90
Urea, concentration, 41, 44
  elimination, 41
Urinary secretion, 62
Urinary tract, 45
Urine
  acidity, 44, 122–129
  camel's, 11
  concentration, 44
    desert mammals, in, 84
    sodium and chloride ions, 88
  jerboa's 13, 42–45, 48–49, 122–131
  man's, 81
  nutrition, 42–45, 48–49, 122–131
  rat's, 13

Vaporization of water, latent heat of, 64
Vapour pressure, atmospheric, 5
  gradient, skin-to-air, 82
Vasoconstriction, 59, 82
Vasodilatation, 59, 82
Vegetation, 2, 6, 7, 24
Vitamins, 85
'Voluntary dehydration', 80, 89

Wadi Digla Desert, 5, 27
Wadi Halfa, 15
Wadi Hawar, 10
Water
  balance, 9, 12, 62, 72–75, 80, 83–85, 107–110
    body weight, 89
    diet, 72–75
    plasma concentration, 89
  conservation mechanisms, 34, 84–85
  content of the body, 45, 110–111
  deficit and dehydration, 88–90
  drinking animals, 9
  economy, 63
  exchanges with the environment, 81
  excretion, 83–84
  extracellular, 83
  gain (income)
    dry and wet diet, 73–74, 130–131
    oxidation (metabolic water), 73–74
    preformed (natural in food), 73–74
  inanition, 62
  intracellular, 83
  loss (expenditure)
    dehydration, 88–90
    dry and wet diet, 73–74, 130–131
    evaporation, 73–74
    feces, 73–74
    sweat, 85–90
    urine, 73–74
  metabolism, 42, 65
  movements under thermal stress, 84
  oxidation (metabolic), of, 10, 12, 65, 73–74
  preformed (natural in food), 9, 12, 73
  routes of elimination, 81
  scarcity, 31, 62, 74
  stored in the body, 73–74
  sub-soil resources, 2
Winds, 2, 22
  khamsins, 4
  simooms, 4
  siroccos, 4
  velocity, 5, 82
Working efficiency, 86

COLLEGE OF THE SEQUOIAS
LIBRARY